Health Informatics

(formerly Computers in Health Care)

Kathryn J. Hannah   Marion J. Ball
Series Editors

T0185170

Springer

New York
Berlin
Heidelberg
Barcelona
Hong Kong
London
Milan
Paris
Singapore
Tokyo

Health Informatics Series
(formerly Computers in Health Care)

Series Editors
Kathryn J. Hannah   Marion J. Ball

(continued after Index)

James G. Anderson   Kenneth W. Goodman

Ethics and Information Technology

A Case-Based Approach to a Health Care System in Transition

Springer

James G. Anderson, PhD
Department of Sociology
  and Anthropology
Purdue University
West Lafayette, IN 47907, USA
andersonj@sri.soc.purdue.edu

Kenneth W. Goodman, PhD
Ethics Programs
University of Miami
Miami, FL 33101, USA
kgoodman@miami.edu

Series Editors:

Kathyrn J. Hannah, PhD, RN
Adjunct Professor, Department
  of Community Health Science
Faculty of Medicine
The University of Calgary
Calgary, Alberta, Canada

Marion J. Ball, EdD
Vice President, Clinical Solutions
Healthlink
2 Hamill Road
Quadrangle 359 West
and
Adjunct Professor
The Johns Hopkins University School
  of Nursing
Baltimore, MD, USA

Cover illustration by Bruno Budrovic © 2002 Images.com, Inc.

Library of Congress Cataloging-in-Publication Data
Anderson, James G.
   Ethics and information technology: a case-based approach to a health care system in
transition / James G. Anderson, Kenneth W. Goodman.
      p. cm.— (Health informatics)
   Includes bibliographical references and index.
   ISBN 978-1-4419-2947-1   e-ISBN 978-0-387-22488-6
   1. Medical informatics—Moral and ethical aspects.   2. Medicine—Information
services—Moral and ethical aspects.   3. Computer networks—Social aspects.   I. Goodman,
Kenneth W., 1954–   II. Title.   III. Series.
R858 .A475 2002
174'.2—dc 21   2001032842

Printed on acid-free paper.

© 2010 Springer-Verlag New York, Inc.
All rights reserved. This work may not be translated or copied in whole or in part without the written permission
of the publisher (Springer-Verlag New York, Inc., 175 Fifth Avenue, New York, NY 10010, USA), except for brief
excerpts in connection with reviews or scholarly analysis. Use in connection with any form of information storage
and retrieval, electronic adaptation, computer software, or by similar or dissimilar methodology now known or
hereafter developed is forbidden.
The use in this publication of trade names, trademarks, service marks, and similar terms, even if they are not
identified as such, is not to be taken as an expression of opinion as to whether or not they are subject to proprietary
rights.
While the advice and information in this book are believed to be true and accurate at the date of going to press,
neither the authors nor the editors nor the publisher can accept any legal responsibility for any errors or omis-
sions that may be made. The publisher makes no warranty, express or implied, with respect to the material contained herein.

Production managed by Michael Koy; manufacturing supervised by Jacqui Ashri.

Printed in the United States of America.

9 8 7 6 5 4 3 2 1

Springer-Verlag   New York   Berlin   Heidelberg
A member of BertelsmannSpringer Science+Business Media GmbH

Series Preface

This series is directed to health care professionals who are leading the transformation of health care by using information and knowledge. Launched in 1988 as Computers in Health Care, the series offers a broad range of titles: some addressed to specific professions such as nursing, medicine, and health administration; others to special areas of practice such as trauma and radiology. Still other books in the series focus on interdisciplinary issues, such as the computer-based patient record, electronic health records, and networked health care systems.

Renamed Health Informatics in 1998 to reflect the rapid evolution in the discipline now known as health informatics, the series will continue to add titles that contribute to the evolution of the field. In the series, eminent experts, serving as editors or authors, offer their accounts of innovations in health informatics. Increasingly, these accounts go beyond hardware and software to address the role of information in influencing the transformation of healthcare delivery systems around the world. The series also increasingly focuses on "peopleware" and the organizational, behavioral, and societal changes that accompany the diffusion of information technology in health services environments.

These changes will shape health services in this new millennium. By making full and creative use of the technology to tame data and to transform information, health informatics will foster the development of the knowledge age in health care. As coeditors, we pledge to support our professional colleagues and the series readers as they share advances in the emerging and exciting field of health informatics.

Kathryn J. Hannah, PhD, RN
Marion J. Ball, EdD

Preface

The rapid growth of computer-based information technology is transforming the delivery of health care. Not only does the new technology affect clinical practices and the delivery of health services, but it also enables consumers to assume more responsibility for their own health care. These developments represent a cultural change in health care that is far reaching.

Health care is information-intensive, and health information has become an important commodity in the marketplace. Health care organizations spend only about 2% of their revenues on information technology, but this percentage is expected to rise to levels more in-line with industries like the financial services industry which invests about 12% of revenues in information technology. Products and services under development run the gamut from electronic medical records to data warehouse applications.

While these technological developments hold considerable promise, they raise social and ethical concerns. Some of these concerns (for example, the debate over proposed federal regulations to protect the privacy of medical records) have become quite controversial. The purpose of this book is to provide an outline of some of the major social and ethical issues raised by applications of computer-based information technology in health care.

This volume is divided into an introduction and six chapters, each corresponding to an ethical issue or domain of interest. Each chapter is further divided into sections, case studies, and guidelines for discussion, as appropriate. The headings should not be taken to exhaust an issue or rigidly plot its boundaries because many issues are related to each other in interesting and sometimes complicated ways. For instance, it seems that questions of privacy and confidentiality, often the first among equals in discussions of ethics and informatics in health care, must be given an extensive chapter. Similarly, the extraordinary growth of the World Wide Web compels us to devote a chapter to suitable cases.

However, notice that confidentiality is an ethical issue and the World Wide Web is what we have called a *domain* of interest. In which chapter, therefore, should we place case studies involving the confidentiality of health information *on the Web*? In some instances, the answer to this question will be that it doesn't matter. The case could go anywhere. In other instances, it makes sense to place the case in one area or another according to its most salient features. Therefore, if the Web shaped the most important features of a case involving confidentiality, that case would be most appropriately located in the Web chapter. Throughout, of course, these decisions required many diverse judgment calls.

Each chapter contains an overview of major issues followed by a collection of actual cases and bibliographic resources designed to enable the reader to gain further insight into the issues raised. The cases reflect social and ethical issues raised by the use of information technology in health care settings.

We have tried to keep the cases as focused as possible. This involved tradeoffs between accessibility, on the one hand, and completeness, on the other. We have most often decided that the virtues of brevity outweigh those of comprehensiveness because this best reflects actual decision-making circumstances. That is, in the health professions only rarely do we make decisions in environments shaped by perfect or complete information. Indeed, it is quite often the case that the most challenging scientific and ethical decisions are those shaped as much by what we do not know as by what we do. Thus, one of the first duties of those using these cases is to identify those areas in which more information is needed. From there, the user also has the opportunity to vary some facts and see what effect this has on subsequent debate and discussion.

Although most of the cases were identified especially for this volume, a few have become "classics" and are well known. The former are included to demonstrate that there is nothing extraordinary about ethics and informatics—issues are everywhere. The latter are included to make plain that some cases have already been noteworthy enough to attract attention and spark debate. Generally, the "classics" are of such importance that it would be an error to miss this opportunity to include them.

Each case—in some instances a cluster of cases—is accompanied by a set of questions for discussion. These questions (some are best thought of as challenges, or even provocations!) are intended to guide readers through what seem to us to be the most important features of the case. Some explicitly propose hypothetical additions to the facts in order to stimulate more effective reflection, analysis, and discussion. One measure of our success will be the extent to which readers end up discussing questions not raised explicitly.

Many of the issues are basic and have been raised in the past; for example, privacy and confidentiality, and conflict of interest. Others, such as licensing and regulation when health care is provided online, are new. The cases are ideal for use in medical education and professional development activities, such as continuing health education, as well as by ethics committees because they are pertinent to the experiences of health care professionals. The source of each case is given, and we encourage readers to consult these sources for more background information.

There are major advantages for using a case-based approach in order to introduce professionals and students to social and ethical issues arising out of informational technology applications in health care. Case discussions permit participants to share opinions based on their various experiences and values, and frequently lead to the identification of issues and the definition of principles to guide practitioners. Case studies can be used to test these principles for comprehensiveness. Each case can be approached from a variety of points of view. Although we have arranged the cases into categories that make sense from our perspective, we invite the reader to reorganize them into personally meaningful categories. For professionals, case studies provide reference points when they experience analogous situations in the performance of their professional duties.

Chapter 1 outlines how ethics can be applied in health informatics and serves as an introduction to the book. There is much evidence that the use of case material to identify ethical issues is a valuable way to train professionals in medicine, nursing, and allied health professions. Moreover, ethics provides tools for addressing these issues and contributes to the development of professional standards.

The second chapter discusses the rise of online health care, including virtual house calls, medical records storage, the sale of organs and pharmaceutical products, and mental health services—in other words, a host of issues, some old, some new. For example, online services that cross state and national boundaries raise issues about regulation and licensing, in addition to concerns about privacy and confidentiality.

The availability of health information online is having major effects on consumers, a topic that is addressed in Chapter 3. It is estimated that 48% of adult Americans, about 97 million people, use the Internet to acquire information, products, and services. Annually, one in three Americans uses the Internet to obtain health information. Access to health information can aid consumers in making informed decisions about health care and in assuming more responsibility for managing their own health. However, though a wealth of health information is available, consumers may also find inaccurate, misleading, unverified, and, sometimes, fraudulent information. Conflict of interest issues are raised by commercially sponsored Web sites that provide health information.

Chapter 4 addresses privacy and confidentiality issues raised by information technology applications in health care. The importance of electronically collecting, storing, analyzing and using health information is undisputed. Consumers need information to make informed choices; physicians need it to provide quality care; and health plans need it to assess outcomes, control costs, and assure quality. Such uses of information technology pose a dilemma, however. How can we provide the required data while at the same time protecting the privacy of patients? Problems arise because of accidental breaches of security and inappropriate use of health information, and in the course of secondary uses of health information by employers, insurance companies and researchers.

Information technology and genetics are likely to have among the greatest effects on health care in the twenty-first century. The use of information technology to acquire, store, manage, analyze, and transmit genetic data presents exciting opportunities and a host of new challenges. Issues raised in Chapter 5 include accuracy and error, appropriate use of genetic information, and privacy and confidentiality.

Clinical information systems provide major benefits in direct support of patient care. These benefits include increased efficiency in managing clinical information, improved quality of care, and cost savings through decision support and improved patient management. However, barriers exist in the implementation and use of clinical information systems. Surveys have indicated that less than 10% of acute care hospitals in the United States have computerized all areas of the hospital. In ambulatory care settings, computer-based patient record systems have been implemented in only about 5% of group practices. Failures are abound in implementing health information systems. Furthermore, some newly implemented systems harm patients. Chapter 6 presents cases that illustrate these issues.

The Internet and the World Wide Web have become essential tools to biomedical researchers. Consumers can seek out and join clinical trials online. Online surveys provide researchers with new means of collecting research data. Researchers also study users of information technology. Chapter 7 addresses social and ethical issues raised by these applications.

This collection of cases reflects health care systems in transition. A major value of the collection is that it identifies important social and ethical issues raised by the introduction of information technology into health care. Our goal is to provide a broad collection of resources which professionals and students might profitably use to identify, discuss, and debate controversies and challenges raised by these developments.

James G. Anderson, PhD
Kenneth W. Goodman, PhD

Acknowledgments

The idea for this book arose out of a collection of cases begun as a project sponsored by the Ethical, Legal, and Social Issues Working Group of the American Medical Informatics Association (AMIA). We are grateful to many members of the association who helped us to identify cases. AMIA provided funding for some of the library work required for the documentation of cases and the preparation of bibliographies. AMIA deserves our first and foremost acknowledgment and thanks for its encouragement of activities that address ethical issues in the profession and, in particular, the effort that led to this book.

Springer-Verlag provided encouragement and financial support for the preparation of the book manuscript. Earlier versions of many of the chapters were published in the journal *MD Computing* with the assistance of Nhora Cortes-Comerer, Executive Editor of Medical Informatics. We also wish to thank Dr. Kathryn J. Hannah and Marion J. Ball, editors of the Health Informatics Series, for their encouragement and support.

We are especially grateful to Marilyn Anderson for her assistance with every aspect of the project. She devoted many hours to identifying and documenting cases, preparing questions for discussion and references for further reading, and overseeing the preparation of the manuscript. Her efforts were essential in making this book a reality. We thank her for her able assistance.

Throughout the text we have identified the sources of cases that have been published. In some instances, professionals who have chosen to remain anonymous contributed cases. We thank them for their contributions. The names of individuals who helped identify cases are listed below in alphabetical order. We wish to express our gratitude to each of them for sharing his or her experience.

Sheri A. Alpert, University of Notre Dame, South Bend, Indiana
Joan Ash, Oregon Health Sciences University, Portland, Oregon
Keith Bauer, Madisonville, Tennessee
John Carpenter, Oregon Health Sciences University, Portland, Oregon
John Christiansen, Counsel, Stoel Rives LLP, Portland, Oregon
T. Hiruki, Oregon Health Sciences University, Portland, Oregon

M. Krall, Oregon Health Sciences University, Portland, Oregon
Armand H. Matheny Antommaria
J.D. Miller
D. Smith, Oregon Health Sciences University, Portland, Oregon
R.H. Strube

Finally, a number of efforts are underway to create codes of ethics and instruments that can be used to rate health-related Web sites. We wish to thank the relevant editors, publishers, and organizations listed below for permission to reprint material included in the appendices. These are:

American Medical Association
Health On the Net Foundation (HON)
Internet Healthcare Coalition
TRUSTe Privacy Program
Hi-Ethics, Health Internet Ethics
Health Summit Working Group

James G. Anderson, PhD
Kenneth W. Goodman, PhD

Contents

1

Introduction: Case Studies in Ethics and Health Informatics

Reprinted with permission from Harley L. Schwardron.

There is nothing like a fledgling science to set our moral teeth on edge. From organ transplantation to assisted reproduction, and from genetic engineering to stem cell therapy, the strivings of scientists and clinicians always seem to be at least a step ahead of ethical analyses. This should occasion no despair but, rather, a sense of excitement and opportunity. The emergence and evolution of a new technology gives us a chance to test old tools and, as necessary, to invent new ones in order to get better moral leverage on the problems at hand. Such tools will inform our decisions, guide our actions, and prepare us for future challenges.

In the management of health information, the use of computers, databases, and networks may serve as a test of our ability to apply moral insight to practical decision making. We can put this in the form of a straightforward question:

How can ethics be of use to health informatics?

The answers will cluster around the following functions:

- Identifying issues and illustrative cases
- Providing tools for addressing and attending to these issues
- Grounding an emerging profession in a body of standards

Issues and Cases

There is plenty of good evidence from clinical nursing, medicine, psychology, and other professions that the identification of ethical issues is a useful way to train newcomers, solve problems, and anticipate (and sometimes prevent) future difficulties. Therefore, among the first tasks in any inquiry is the identification of which topics, subjects, and issues are valid components of the ethical enquiry, that is which have an identifiable relationship to the larger domain of ethics. Another way of putting this is to ask "Is that an *ethical* issue or another kind of issue?" For instance, an issue such as the best way to display lab results at bedside terminals, or which database software best facilitates outcomes research, is likely to be a technical issue, a financial issue, or a workflow process convenience issue, rather than an ethical one.

An ethical issue is one that embodies questions about whether an action is good or bad, right or wrong, appropriate or inappropriate, praiseworthy or blameworthy. (To be sure, there are nonethical senses of "good," "inappropriate," and "blameworthy," and so forth, and many interesting philosophical questions surround these uses.)

So, if bioethics, that branch of applied ethics that deals with the life sciences and health professions, is to be used to good effect in health informatics, it must help us identify ethical issues that arise in the field. There is good reason to believe that bioethics is able to provide this service.

For instance, personal health information should generally be kept secret or held in confidence. Confidentiality becomes an ethical issue in health informatics both for traditional reasons (trust in disclosure, fear of stigma and discrimination, etc.) and contemporary ones (e.g., computers and networks broaden access to information, permit links and associations not previously possible). Once confidentiality has been identified as an ethical issue in health informatics, we need to attend to the following questions, among others [1]:

- To what extent should confidentiality be safeguarded?

- How should we distinguish between and balance legitimate access and illegitimate access?
- Under what circumstances may confidentiality be violated or breached?

Almost anyone who works in health information management will know of instances in which confidentiality has been challenged. Identification of a case or case study attaches a story to these instances. One goal of this book is to tell and comment on some of these stories. The chapter headings and sub-headings reflect issues that arise in health informatics.

Thus equipped with issues and cases we are well on the way to doing ethics in health informatics. To *do ethics* is to apply critical reason to arguments made by various parties in the case studies. Suppose it is argued that all hospital employees should have access to all patient records. A counter-response is that such broad access would unnecessarily violate confidentiality and not improve patient care. The process of agrument and counter-argument continues until a conclusion is reached. Such exchanges can then be used to craft appropriate institutional policies.

While confidentiality is often the first ethical issue identified in health informatics, others are just as interesting and important (references 2 and 3 contain sustained treatments of all these issues). Some relate to challenges posed by the use of decision support systems and prognostic scoring systems:

Appropriate use: What is an acceptable use of a diagnostic expert system? Should such systems be used to render diagnoses or to assist competent humans in identifying maladies? If there is evidence that such systems can help arrive at more accurate diagnoses, is it blameworthy if clinicians do not use them? Is it permissible to use a computer system that calculates prognoses to guide critical care units in decisions to terminate care?

Appropriate users: Who should use these systems? Must they be physicians? Physicians and nurses? Students? Lay people? And how much education is appropriate before a person uses a computer and software for decision support?

Accountability and responsibility: Who is to blame if something goes wrong? Because many people and entities are involved in designing, building, updating, and using decision support systems, it can be a difficult challenge to determine where responsibility should be assigned.

Other issues concern the World Wide Web, system evaluation, research (including outcomes and quality assessment research) and the patient-professional relationship. Clearly, the intersection of ethics and health informatics is a rich and varied field. As with other aspects of professional life, ethical issues are best addressed by educational programs and better communication. Bioethics provides precisely the conceptual tools needed for such efforts. Case studies provide the grist for this conceptual mill.

Practical Tools

Bioethics can serve informatics by providing the means to address ethical issues and resolve ethical conflicts. Once we begin the process of identifying ethical issues and salient cases, we should consider how to apply any new understanding and insight to institutional conflicts, controversies, and challenges. It is one thing to say that confidentiality is an ethical issue and to identify instances in which confidentiality was violated or imperiled; it is quite another to improve policies and procedures and to apply the insights of applied ethics to ongoing cases.

Suppose someone expresses discomfort at the prospect of using patient records without patient consent for outcomes studies or quality improvement. A first step in any sound ethics process is to gather adequate information, that is, to flesh-out the case. We need to know if the patient records retain any unique identifiers or if they will be anonymized. We need to know if it is possible to obtain consent for such use of personal information, and to know the consequences for the studies if dissenters' data are excluded. We need to know what benefits and harms will follow if the studies are completed as contemplated as well as the benefits and harms if they are not completed.

Some institutions successfully approach this issue by establishing policies that require the anonymization of information and the disclosure of outcomes and quality research on admission.

One way to consider how best to proceed in these contexts is to seek answers to the question, "How can this activity be *ethically optimized*?" Few policies—indeed, few actions—are ethically unambiguous. We are often faced with decisions that require us to balance harms and benefits; it is in the course of ethical scrutiny that we can make the best decision given the facts of the case.

To be sure, there are many clear cases. Suppose someone proposes that patient information be sold to a commercial venture, and that the proceeds be placed in the general fund to reward corporate stockholders. In such a case, the harms are so great (violation of confidentiality, erosion of trust, risk of discrimination, etc.) and the benefits so meager (better returns for investors) that it is easy to identify the right thing to do.

It is important to note that the Joint Commission on Accreditation of Health Care Organizations [4] requires all accredited bodies to have a mechanism for addressing ethical issues. Many, perhaps most, institutions comply by establishing ethics committees. Alas, these committees are largely ignorant of the ethical issues that arise in health informatics. Health information professionals should consider bringing appropriate issues in informatics to the attention of ethics committees. An institution could thereby build an ethics committee better able to address issues in health informatics and hence one of greater and broader service.

Any institutional mechanism for addressing ethical issues now needs to fulfill three functions: education, policy development and review, and consultation. So another way bioethics can serve health informatics (and improve institutional ethics efforts in general) is to inform and guide policies for information management. Policies that lay out appropriate access to online patient information should be guided by concerns for both confidentiality and ready access for appropriate persons. An ethically optimized policy embodies a practical approach to real-world challenges, and navigates by the light of broadly shared moral values. Ethics committees are one mechanism for crafting such policies.

Health information professionals thus have an opportunity to raise the standards by which ethical issues are addressed and conflicts resolved. By using the bioethics expertise that might be available through an ethics committee, health information professionals foster relationships that can, in principle at least, both educate the committee and improve decision making elsewhere in the institution.

Professional Grounding and Standards

A third way that bioethics can be of use to health informatics is by contributing to the development of professional standards. Part of being a professional is being a member of a group that hews to practice standards. There are different types of standards: quality, practice, education, and so forth. We should be prepared to recognize a standard of ethics for professionals. What this ethical standard entails can range from the articulation of core values and duties to a readiness to commit to continuing education in order to remain up to date on current issues, methods, and controversies.

Codes of ethics are one popular way to lay out the ethical foundation of a profession, and a number of codes have been proposed for health information professionals [5]. However, the main value of a code should be seen to be educational. In isolation, ethics codes or guidelines will be relegated to the shelf, along with so many other well-intentioned documents that few read, understand, or follow. But an ethically optimized code or set of guidelines can be of great utility in an education program by demonstrating that the discipline enjoys a grounding in values and standards.

An ethics faculty can lead courses, seminars, workshops, in-service presentations and other activities. Sometimes, the use of case studies, a standard method of ethics training in such areas as nursing, medicine, and public health, is just the way to engage professionals and students in the formation of approaches to the ethical conflicts and challenges of daily practice [6–8].

The phrase *ethics faculty* is used to underscore the importance of identifying colleagues who are competent for these tasks. Bioethics has become something of a growth industry, with scholars, clinicians, administrators, lawyers, and others taking an interest. The fact that a person thinks bioethics is important, interesting, or engaging does not, however, render that person competent to teach it. So if bioethics is to provide the high quality of service to health informatics being proposed here, it is essential to identify and/or train people competent to carry out the tasks of education is bioethics. Ethics is itself a profession, with evolving standards for competence, and one can do real mischief in ignorance of its practical, theoretical and pedagogic underpinnings.

Synthesis

The Information Age offers extraordinary opportunities for society, not least to those components concerned with providing health care. The rapid—no, the *dizzying*—growth of health informatics provides an example of how these opportunities are not unidimensional but, instead, present challenges, engender conflicts, and foster controversies.

Fortunately, tools are available to enable health professionals to address those challenges, manage those conflicts, and learn from those controversies. How? By identifying issues and illustrative cases, by applying critical reasoning in grappling with these issues, and by grounding the health informatics profession in a body of standards and values.

This is not a feel-good exercise, a bit of busy work to impress evaluators, or a scholarly debate removed from practical concerns. It is an effort, in the service of better health care, to manage and learn from the exciting and difficult challenges that emerge at the intersection of computing, healthcare and ethics.

References

1. National Research Council. *For the record: protecting electronic health information.* Washington, DC: National Academy Press. 1997.
2. Miller RA, Schaffner KF, Meisel A. Ethical and legal issues related to the use of computer programs in clinical medicine. *Ann Intern Med.* 1985;102:529–536.
3. Goodman KW (ed). *Ethics, computing, and medicine: informatics and the transformation of health care.* Cambridge: Cambridge University Press, 1998.
4. Joint Commission on Accreditation of Health Care Organizations. *Ethical issues and patient rights: across the continuum of care.* Oakbrook Terrace, IL: Joint Commission on Accreditation of Health Care Organizations, 1998.
5. Kluge EHW. Fostering a security culture: a model code of ethics for health information professionals. *Int J Med Inf.* 1998;49:105–110.

6. Coughlin S, Soskolne C, Goodman K. *Case studies in public health ethics*. Washington, DC: American Public Health Association, 1997.
7. Pence GE. *Classic cases in medical ethics*. New York: McGraw-Hill, 1995.
8. Veatch RM, Flack HE. *Case studies in allied health ethics*. Upper Saddle River, NJ: Prentice Hall, 1997.

Further Readings

Allaërt FA, Dusserre L. Télémédecine et responsabilité médicale. *Arch Anat Cytol Pathol.* 1995;43:200–205.

Barhyte DY. Ethical issues in automating nursing personnel data. *Comput Nurs.* 1987;5;171–174.

Brody BA. The ethics of using ICU scoring systems in individual patient management. *Probl Crit Care.* 1989;3:662–670.

Colby KM. Ethics of computer assisted psychotherapy. *Prof Psychol Res Pract.* 1986;19:286–289.

Cowen JS, Kelley MA. Error and bias in using predictive scoring systems. *Crit Care Clin.* 1994;10:53–72.

Davies BL. A discussion of safety issues for medical robots. In: Taylor RH, Lavallee S, Burdea GC, Mosges R, eds. *Computer-integrated surgery: technology and clinical applications.* Cambridge, MA: MIT Press; 1996:287–296.

de Dombal FT. Ethical considerations concerning computers in medicine in the 1980s. *J Med Ethics.* 1987;3:179–184.

Denning DE, Lin HS, eds. *Rights and responsibilities of participants in networked communities.* Washington, D.C.: National Academy Press; 1994.

Doroszewski J. Ethical and methodological aspects of medical computer data bases and knowledge bases. *Theor Med.* 1988;9:117–128.

Ford BD. Ethical and professional issues in computer assisted therapy. *Comput Hum Behav.* 1993;9:387–400.

Gamerman GE. FDA regulation of biomedical software. In: Frisse ME, ed. *Proceedings of the sixteenth annual symposium on computer applications in medical care.* New York: McGraw-Hill; 1992:745–749

Goodman KW. Monitoring ethics. *Physicians Comput.* 1993;10:10–12.

Goodman, KW. Critical care computing: outcomes, confidentiality and appropriate use. *Crit Care Clin.* 1996;12:109–122.

Goodman KW. Ethical and legal issues in use of decision support systems. In: E. Berner, ed. *Decision support systems.* New York: Springer Verlag, 1998: 217–233.

Goodman KW. Health informatics and the hospital ethics committee. *MD Comput.* 1999;16:17–20.

Goodman, K, Miller R. Ethics and health informatics: users, standards and outcomes. In EH Shortliffe, LE Perreault, G Wiederhold, LM Fagan, eds. Medical informatics: computer applications in health care and biomedicine, 2nd ed. New York: Springer-Verlag, 2001:257–281.

Kessler DA, Pape SM, Sundwall DN. The federal regulation of medical devices. *N Engl J Med.* 1987;317:357–366.

Kluge EH. Advanced patient records: some ethical and legal considerations touching medical information space. *Methods Inform Med.* 1993;32:95–103.

Miller RA. Why the standard view is standard: people, not machines, understand patients' problems. *J Med Philos.* 1990;15:581–591.

Moor J. Are there decisions computers should never make? *Nat Syst.* 1979;1:217–229.

Oliver AJ. Internet pharmacies: regulation of a growing industry. *J Law Med Ethics.* 2000;28:98–101

Woolery LK. Professional standards and ethical dilemmas in nursing information systems. *J Nurs Admin.* 1990;20:50–53.

2

The Business of Cyber Health Care

"Well, *www.what'swrongwithme?.com* says it's just a virus, but I came to you for a second opinion."

Reprinted with permission from Bunny Hoest.

Along with just about everything else being sold on the Internet these days, health care services are a hot commodity. As more and more consumers learn that health information and medical services are only a mouse click away, demand is surging—and health-based Web sites are continually popping up to meet that demand. A Louis Harris poll estimated that 70 million Americans sought health-related information online in 1999 [1]. That number is expected to increase, leading to a $370 billion online health care industry by the year 2004. A survey of 71 health care companies by Forrester Research estimated that by 2004 consumers will use the Internet to purchase $15 billion in prescription drugs, $1.9 billion in over the counter nonprescription drugs, $3.3 billion in alternative health cures, and almost $1 billion in health and beauty aids [2].

As with most new ways of doing business, the explosion of Web sites offering health services is raising a number of important issues, particularly in the area of safety: Can advice and diagnoses really be effective without any physical contact between online clinicians and patients? What is the level of accuracy and reliability of the medical information these sites dispense to consumers? What is the potential for conflicts of interest (see Chapter 3)? How secure is patient data that is collected, stored, and accessed electronically (see Chapters 4 and 5)? What are the risks to patients and to the health care organizations providing care when health information systems are not adequately evaluated and implemented (see Chapter 6)? How can the rights of subjects of biomedical research that involves the Internet and the Web be protected (see Chapter 7)?

Despite the ongoing ethical debates, the moneymaking potential of online health care has attracted the attention—and the financial backing—of a number of major corporations. American Online, the largest Internet service provider, in 1999 put its resources behind former US Health Secretary Dr. C. Everett Koop's Web site (the site later underwent major revisions amidst allegations of conflict of interest); NBC purchased a stake in iVillage, which sponsored the Better Health Web site. Medscape, a company that provides physicians with continually updated medical information via the Web, joined with CBS to offer consumer health information as well. Time Warner's CNN News Group bought an equity share in WebMD, which merged with Healtheon, another Internet health care company. Dupont also invested in WebMD by offering free five-year subscriptions to the information service for 200,000 physicians [3].

The Online Medical Marketplace

The World Wide Web allows consumers to seek all manner of health services: health information, social support, a diagnosis, a prescription, surgical and mental health services, a virtual house call, and information about clinical trials. For example, NetWellness is a major Internet consumer health information site. Its "Ask an Expert" service features 200 health professionals from medical centers who answer consumer questions on more than 40 topics. It also provides links to other health Web sites that the Health Summit Working Group has evaluated using its Criteria for Assessing the Quality of Health Information on the Internet (Appendix 6).

A number of other Web sites offer consumer health information. AmericasDoctor.com provides a free service called "Ask the Doc for," where consumers can contact an Internet physician to ask health-related questions. The doctors, however, will not diagnose the patient's illness or prescribe medications. Mediconsult.com charges a fee of $195 to provide personalized information on alternative treatments. Consumers submit a detailed health history and receive reports suggesting the best treatment option. WebMD.com offers online group discussions moderated by medical experts while Cyberdocs.com provides "virtual house calls" for a fee of $50 to $100. The patient/consumer can talk electronically to a board-certified specialist, who is authorized to diagnose minor illnesses and prescribe medications for them [4].

Consumers can also shop for surgical procedures on the Internet [5]. Consumers enter their medical profile and list the surgical procedure they want. Physicians and dentists can bid on the procedure and negotiate their fee with the consumer for elective procedures such as face lifts, breast augmentations, laser vision correction, and dental braces.

The Internet has also made possible the virtual house call [6]. Patients with chronic diseases such as diabetes, heart disease, and asthma can be monitored in their homes or offices. Readings such as blood glucose levels can be sent to a clinic with a personal computer over the Internet. Doctors and nurses can monitor the patient's data and send back evaluations and treatment instructions.

Several Web sites contain information on clinical trials. These include drkoop.com, medtrial.com, clinicaltrials.com, nci.nih.gov, and CenterWatch.com. Patients can enter a description of their illness and search for a study they might wish to join.

Online Medical Records

Consumers can also post their medical records on the Web. PersonalMD.com is one such site where consumers can maintain their health history, X-ray and EKG reports, emergency medical data, problem list, and current medications. In an emergency, the patient or healthcare provider can access this information [7].

In the near future, the HIV Treatment Data Project will provide a Web site that will publish accounts of patients' experiences with antiretroviral drugs [8]. The project is a collaborative effort among AIDS advocacy organizations, academia, and industry. It will function as a repository for HIV-infected persons to record the drugs they are taking, their progress, and how they feel. Supporters maintain that the project will make it possible to collect national data on antiretroviral drugs so that clinicians can make decisions based on the experiences of a large number of patients. The Web site will also provide feedback to patients and healthcare providers on the best practices for treating HIV.

Pharmaceutical Products

There has been a rapid growth in online demand for pharmaceutical products. Some sites, such as Cyberpharmacy.com, Drugstore.com, and Soma.com, only sell drugs after a doctor has seen the patient and signed the prescription. Cyberexpress.com, for example, asks the patient to answer an online questionnaire whereupon a medical consultant (an MD) provides the prescription for the drug [9, 10]. Other sites dispense drugs to patients without a prescription.

Psychological Services

Psychological services are also moving behavioral health onto the Web. Concerned Counseling is a network of therapists who use the Internet to provide counseling to individuals and families [11]. Counselors are available for chatroom consultation or the patient can send an e-mail message. Chatroom sessions cost $45 for up to 30 minutes. E-mail consultations cost $30 for each response.

Unresolved Issues

While the use of the Internet is transforming the way health care is provided, there are a number of issues that need to be resolved, insurance is one of the major ones. Many of the health services provided via the Internet are not covered by third-party reimbursement. Another problem is the limited ability of consumers to determine the accuracy of Web-based health information. One study looked at information on childhood diarrhea from 60 sites and compared it with official information from the American Academy of Pediatrics [12]. Eighty percent of those sites contained inaccurate information. Far worse, the Federal Trade Commission identified hundreds of Web sites promoting phony cures for more than 30 illnesses such as HIV/AIDS, multiple sclerosis, and cancer [13].

Provider/patient anonymity is another problem. Consumers who seek personal advice online often have no way of checking the credentials of the person who is providing the information. Another area of concern is that physicians who diagnose and prescribe over the Internet do so without physical examinations and diagnostic tests. This raises a major safety issue: Prescribing medications without the benefit of a physical exam raises the risk of adverse drug events. The chances of a bad drug reaction are further increased by consumer access to Internet pharmacies that provide drugs that haven't been adequately tested and approved. The American Medical Association (AMA) has condemned online practices by physicians, but local or regional medical boards do not have the staff resources necessary to investigate the large number of Internet sites that provide diagnoses and prescriptions [14].

In the future, there will likely be more regulation of the Internet health care industry as legislative bodies consider a broad array of laws [15]. The National Association of Boards of Pharmacy awards a seal of approval to Web sites that meet its criteria for dispensing prescription medications. At least five states have begun investigations of Web sites they suspect are violating state medical practice laws or not adhering to accepted standards of care. In one instance, an Ohio grand jury indicted a physician for dispensing drugs over the Internet without appropriate medical consultation.

Conflicts of Interest

Consumers who seek advice on treatment and medications on the web are generally unaware that drug companies have a major financial stake in many of the health Web sites. On August 16, 2000, for example, Newsweek.com reported that NicoDerm was sponsoring the drkoop.com Web site section related to the treatment of tobacco addiction [4]. The article also reported that DuPont had a major investment in WebMD and that the company had exclusive rights to provide information on nutritional supplements. As a result of such observations and publicity, Web sites are beginning to alter their arrangements with sponsoring companies. Consumers need to be aware, however, that potential conflicts of interest may influence sites to provide selective or inappropriate information to the consumer who seeks medical advice.

Another potential conflict of interest concerns Web sites that promote health centers in exchange for commissions and fees. For example, healthcare institutions listed on the drkoop.com site were described as "...the most innovative advanced health care institutions across the country." The $40,000 fee each institution paid for that praise was not mentioned, nor did the site acknowledge this until it encountered public scrutiny and criticism [16].

Confidentiality and Reliability

Placing personal medical data on a Web site raises important issues of privacy and confidentiality. In one instance, the sexual and medical histories of patients being treated by a psychiatrist (who was also a certified sex therapist) were unintentionally posted on a public Internet site [17]. Some sites, such as PersonalMD, use encryption to guard against unauthorized access to personal data. However, it is impossible to provide absolute security, and it is often unclear whether sites that warehouse medical histories sell their clients' data to other businesses. *The Washington Post* has disclosed that drug companies, including Glaxo Wellcome, Warner-Lambert, Merck, Biogen, and Hoffman-LaRoche, paid drugstore megachain CVS for pharmacy prescription data that they then used to market their drugs directly to patients. As a result, a class-action suit challenging this practice was filed [18].

Another problem that arises is the reliability of the medical information stored on the Web. When patients are responsible for posting their own medical records, it is difficult to insure accuracy and completeness. Patients may be reluctant to include certain data such as their HIV status or psychiatric history. They may also fail to update data concerning recent health problems and medications. Consequently, when emergencies arise, healthcare providers may be basing their decisions on unreliable or out-of-date medical information.

Medical research may also be adversely affected by participants who share information online, since such an exchange among patients enrolled in a clinical trial makes it possible to determine who is receiving a placebo and who is actually receiving an experimental drug. This actually occurred in a study of amyotrophic lateral sclerosis (ALS) during clinical trials of a new pharmaceutical, Neuronton [19]. Investigators worried that patients receiving a placebo may have visited other doctors to request a prescription for the drug.

Buyer Beware

The market for consumer health information on the World Wide Web is growing larger and more diverse. Investment in this fledgling industry is increasing rapidly, and buyers need to keep a sharp eye. These enterprises can provide timely, high-quality medical services and give consumers greater control over their own medical records and healthcare, but they can also be dispensaries of incomplete or even harmful information. While Web sites sponsored by the National Institutes of Health, the National Library of Medicine, and some university medical centers provide information based on peer-reviewed journals and advice from professionals with impeccable credentials, other sites function as advertisements for health products and services or promote unproven remedies and therapies. Some sites may sell information about their subscribers to drug companies or to commercial providers of products and services. For people who take advantage of this new medical marketplace, the advice is the same as it ever was: *caveat emptor*.

Case Studies

Health Services Online

Case 2.1: Shopping on the Web for Surgical Options

Description

Consumers can shop on the Web for surgical procedures using Web sites permitting consumers to compare physicians' bids for surgery. Patients log on to such a Web site, enter medical profiles, and list what surgical procedure they want. The online service matches the requests with participating doctors who submit proposals including costs. The patient can review the bids and select a physician to perform the surgery.

The PatientWise site accepts patient requests for 100 procedures including hip replacement, heart surgery, and brain surgery. Medicine Online Inc offers 36 procedures including breast augmentation, liposuction, dental surgery, and laser corrective eye surgery. In both case, medical providers post the services they offer and their charges. Examples of the charges posted on the site are: $4495 for laser vision correction surgery; $70 for nutrition counseling; and $68 for one pair of soft daily wear contact lens. Patients can then bid, and individual doctors and patients can negotiate. If a procedure is not listed on the Web site, the patient can post a request and wait for a provider to respond. PatientWise claims that it verifies all patient and doctor information. However, some sites have a disclaimer saying they cannot verify physicians' credentials. Some sites post information about the quality of care based on patient evaluation. This information might be made available from Healthgrades.com, an online service that rates medical providers.

Source: Naujeck, A. Consumers shopping on Web for surgical options. The Associated Press, May 12, 2000. Available at http://acmi.canoe.ca/Health0005/12_surgery.html. Accessed Sept. 30, 2001.

Questions for Discussion

1. Is it appropriate to treat medical care as a commodity like other goods and services? Why or why not? If you think it is appropriate, are your reasons professional, esthetic, or ethical?
2. What kinds of risks are consumers taking by shopping on the Internet for the cheapest provider of medical services? Should they be allowed to take such risks if they want?
3. What kinds of guarantees should the companies which provide online medical services be required to supply that the providers are qualified to perform the services they offer online? Why not let consumers "pay their money and take their chances?"
4. Is it unethical for physicians to bid on services without first seeing the patient in person? Why, or why not? Indeed, is it unethical to bid on services in the first place? Why, or why not? Note that managed care often involves individuals, practices, and institutions bidding to provide health services for groups of people whom they often have not met in person.

Case 2.2: Let the Buyer Beware

Description

Plastic surgery is highly advertised on the internet. One woman went to a surgeon who advertised heavily on the internet for blepharoplasty or plastic surgery on eyelids. While he claimed to be a board certified plastic surgeon, he actually was an oral surgeon. After the operation, the woman required a corrective skin graft.

Source: Kolata, G. Web research transforms visit to the doctor. *The New York Times.*
    March 6, 2000: A1, A18.

Questions for Discussion

1. How can patients determine the credentials of health care providers who
   advertise on the Internet? Indeed, does online advertising raise issues
   different than traditional print-based advertising?
2. How should medical professional societies sanction physicians who mis-
   represent their skills, or should they? Should society punish online advertisers of medical procedures they are unqualified to perform?

Case 2.3: Virtual Doctors

Description

Virtual Doctors, a former Web site based in New Jersey, invited consumers to fill
out an online consultation form and a questionnaire about symptoms. Visitors to
the Web site then supplied a Visa or MasterCard number for payment. Dr B's name
appeared at the top of the Web site's home page. No information about Dr B's
specialty appeared on the Web site. Dr B is an ear, nose, and throat specialist.
Source: Engstrom P, Brown MS. Electronic house calls: New rules, new roles as
    healers swarm the Net. *Medicine on the Net* (serial online). 1996;2:1–5 Available at:
    http://www.mednet-i.com. Accessed Dec. 15, 1996.

Questions for Discussion

1. How will the "here today, gone tomorrow" characteristics of Web pages
   shape consumers' relationships with physicians online?
2. What can happen to confidential messages, medical histories, interview
   notes, and other sensitive information, such as Social Security numbers
   and credit card numbers, that practitioners and consumers exchange
   online? What should happen to this information?
3. Does a practitioner have an obligation to answer a health or medical
   question in a timely fashion online? What is "a timely fashion"?
4. Should physicians who give advice over the Internet be required to disclose their medical qualifications? Should they have training in Web-
   based practice? What should that consist of?
5. How can consumers of online health care information be assured that the
   advice they receive is objective and not an effort to promote particular
   products and services in which the physician has a commercial interest?

Case 2.4: Online Doctor Visits

Description

Dr F, an emergency medicine physician in Calvert County, Md, recently carried on a dialogue over the Internet with a patient in Houston, Tex. The

patient was concerned about a fluttering sensation in his chest. The physician asked, "Does the fluttering start suddenly and end suddenly or does it sort of creep up on you?" The patient responded that it "creeps up" on him. After about 6 minutes of dialogue, the physician advised the patient that he wasn't entirely certain, but that it sounded as if the condition was not dangerous. He added that it would be appropriate for the patient to visit his physician for a better diagnosis.

Dr F has contracted with AmericasDoctor.com, a Web company with headquarters in Owings Mills, Md, that sponsors the Ask-the-Doctor program. As many as 20 physicians at 2 call centers respond to questions posed by anonymous consumers online. The service is free to consumers. The company receives its revenue from hospital systems whose names and services are displayed on the site once the consumer registers and provides a zip code. At the beginning of each dialogue between a physician and consumer, the Web site displays the following disclaimer: "We do not provide medical diagnosis or treatment advice."
Source: Brown D. Log on and say 'Ahhh'. *Washington Post*, August 22, 1999: A01.

Questions for Discussion

1. In an online dialogue between the doctor and a consumer like this one the physician does not have an electrocardiogram, a medical history, or any other diagnostic information. Is it appropriate for the physician to provide the patient with medical advice? What kind of information would be sufficient?
2. Does this kind of online session fit the definition of *medical practice*? If not, what is missing?
3. Does it constitute a "medical consultation?" Is it merely a casual conversation between the doctor and the consumer? If so, what are the doctor's responsibilities?
4. Is there a danger that inappropriate or inaccurate medical information or advice could result in inappropriate treatment or a delay in treatment? If so, what kind of disclosure is needed for patient participants?

Case 2.5: Litigious Patients

Description

Physicians and hospitals can use a computer service to identify patients who have filed malpractice or other types of civil suits against a healthcare provider. The service, Physician's Alert located in Chicago, Ill, was developed by the Los Angeles County Medical Association. The software has been adopted by the Colorado Medical Society and was to be used by health care providers in a number of other states. Since the database is cumulative, physicians can obtain information about patients who have moved from other cities and states.

In announcing the availability of the service to physicians, the president of the Los Angeles Medical Association stated that the ever-present threat of malpractice suits "... makes it very difficult to practice medicine." Health care providers who use the service can legally turn away patients or refuse to perform specific procedures on patients whose names are included in the database. According to the president of Physician's Alert, most physicians who subscribe to the service use it to check up on patients for elective and scheduled procedures and do not use it for emergency patients.

Source: Program alerts MDs to "lawsuit prone" patients. *National Report Computers and Health.* 1985;6:1–4.

Questions for Discussion

1. Is it ethical for physicians to refuse to provide care or to perform a certain procedure on a patient whose name is included in the Physician's Alert database? Why, or why not?
2. Is it ethical for physicians to have access to the Physician's Alert database when patients do not have access to data about physicians who have been subject to malpractice or civil suits? Why, or why not?
3. Does the existence and use of the Physician's Alert service create or worsen patient and doctor mistrust? Why, or why not?

Case 2.6: Virtual Health Care Systems

Description

Vivius and HealtheCare, 2 Minneapolis-based companies, are planning to sell a complete line of health insurance products online to employer groups. Employers will provide the funding by paying a defined contribution for employees. Providers can set their own charges for their services, to be paid monthly. Employees can pick individual providers and essentially build their own health care system from a list of providers posted on the company's Web site.

Vivius divides health care providers into 20 categories. These include primary care physicians, 14 specialty groups, hospitals, laboratories, and pharmacies. They use an actuary to translate the fees of the care providers into a flat monthly rate for each type of specialty. Physicians are permitted to raise or lower their rates for new customers and alter rates for existing customers after 1 year. Vivius charges higher rates for patient groups that generally have higher medical bills, such as women in their childbearing years and people over 65. Employees pay extra if their choices of providers cost more than the amount that is in the employer's fund. If they choose lower-charging providers, they can apply any savings to other health-related services including health club membership.

Source: Page L. An insurance alternative: Firms selling e-health care. *Am Med News.* 2000;43:1–4.

Questions for Discussion

1. Under what circumstances are Web-based insurance sales appropriate for selecting health coverage? What makes the Web different than other media for insurance product description and registration?
2. Given the high stakes associated with selecting health insurance, how should employers and employees balance the benefits of online communication and decision making with the risks of (potentially) reducing staff support?

Case 2.7: Online Coverage for Uninsured Consumers

Description

Healtheon and Alternative Technology Resources (ATR) plan to create an online system for uninsured consumers to schedule appointments with physicians. Consumers who join the plan will pay a $10 fee and will pay for services with credit cards. Healtheon intends to recruit 300,000 physicians from the ranks of those discontented with health maintenance organizations and managed care organizations. The physicians, in turn, are supposed to provide health services to these consumers at a 15% to 50% discount.

Source: Health care for the wired and uninsured. *The Industry Standard*. October 11, 1999: 47.

Questions for Discussion

1. Is this program likely to appeal primarily to small companies that cannot afford to provide health insurance for their employees? Is this an appropriate solution to the problem?
2. In this system, who will pay for an employee's health care if he or she experiences a catastrophic illness or accident that exceeds his or her credit limits? Who should? Why?
3. Is the plan beyond the reach of most working poor? Does that make it unethical? Why, or why not?

Case 2.8: Smart Cards

Description

A regional medical center in Forida has started using a smart-card system to help improve emergency care and other services. They plan to give smart cards to about 12,000 people, 11,000 of whom are seniors who belong to the Prestige 55 program. The program enables people who are 55 and older who have the smart-cards to register for free and discounted diagnostic services, wellness classes, and other health-related services.

The smart-card is like a credit card with a computer chip that stores patient demographic and medical information such as medications and allergies. Members of the Prestige 55 program pay an $8 annual membership fee for the smart-card program. The hospital plans to expand the use of smart cards to patient registration for inpatient and outpatient services.

Source: Chin, T. Florida hospital deals out smart cards to expedite care. *Am Med News.* April 2000: 1, 37.

Questions for Discussion

1. Is there a danger that physicians using the cards will make decisions based on out-of-date information? If so, how can the information stored on the card be kept up-to-date?
2. Are there new or special risks that arise if cards are lost or stolen? If a card is lost, can this be allowed to cause delays of service?

Case 2.9: An Electronic Soapbox

Description

A hospital in Ohio decided not to renew a contract with 2 physicians who formed a cardiac practice. The physicians had an exclusive contract to perform cardiac surgery at the hospital. The hospital cited disputes at the hospital between one of the 2 physicians and his competitors.

The physician subsequently launched a public relations campaign to pressure the hospital to reverse its decision, hiring a public relations firm and setting up a Web site to protest the hospital's decision and to recruit public support for his campaign against the hospital. To convey the point he was making, he named the Web site www.patientsshouldchoose.com. He registered the domain name, advertised it in the local papers, and registered the Web site with several Internet search engines. When the transcript of a court hearing involving his request for an extension of a temporary restraining order against the hospital became available, he posted the entire document online. In addition, he posted letters of support from patients, sent mass mailings to patients and physicians, and recruited demonstrators to carry signs and march in protest in front of the hospital.

Source: Chin T. Internet offers an electronic soapbox. *Am Med News.* July 31, 2000.

Questions for Discussion

1. Under what circumstances is it acceptable for health professionals to use the World Wide Web to publicize disagreements? Are there good reasons to suggest that health professionals have different or fewer free speech rights than others?
2. Under what circumstances, if any, is it appropriate for a physician to ask the public to put pressure on a hospital to support his claims?
3. How can the public determine the validity of the claims that are made on

such a Web site? Indeed, do you think ordinary people have the ability or inclination to sort out this kind of dispute?

Virtual House Calls

Case 2.10: Cyber Home Health Care

Description

The new health care technologies can provide assistance to family members who are taking care of frail elders. Telemedicine can link the elder and family caregivers more closely with health care providers and other sources of support. Health care workers can check vital signs and weight, examine wounds or injuries that are healing, and oversee the use of medications, using a video camera and electronic monitoring. The Internet can be used to contact health care providers for information; to set up appointments; and to arrange for transportation, meals, and other home services.

Thus, using telemedicine technology including live video and audio, stethoscope, and blood-pressure monitors, a home-health-services provider in New Jersey has set up a diagnostic station in the home of an 85-year-old man with serious heart problems. Health care workers can check on vital signs, view the patient, and oversee the use of medications through videoconferencing and electronic monitoring. Patients can use e-mail to ask questions or to set up appointments. The system provides support to patient and family caregivers, and it warns of impending medical crises.

Source: Shellenbarger S. Taking care of parents: New high-tech links can offer some relief. *The Wall Street Journal*. March 8, 2000: B1.

Questions for Discussion

1. Is there enough evidence to support replacement of standard care strategies with telehealth services? What is the evidence for and against?
2. Does the lack of direct personnel contact between patient and provider diminish the quality of care? In what ways?
3. Medicare and Medicaid do not cover many aspects of telemedicine. Should these programs fully cover health care delivered by telemedicine? Why? Why not?
4. Who should pay for home equipment costs? Patients? Providers? Insurance companies? Under what circumstances?

Case 2.11: The Virtual House Call

Description

Ms D checks her blood sugar while she is in her office. The 46-year-old pricks her finger and squeezes a drop of blood onto a glucose meter. The meter is hooked up to her personal computer by a cable, enabeling her to send the blood sugar level reading over the Internet to Dr P and his staff at a clinic in Atherton, Calif.

Dr P and a group of investors started DiabetesWell.com in October, 1999. To use the service, patients with diabetes register with the Web site and contract to pay $19.95 per month. They then complete a questionnaire that asks for information about their medical history, symptoms, habits, and goals. Each new patient is assigned to a nurse, who keeps in contact by e-mail. The staff will eventually consist of a nurse for every 500 patients and a physician for every 20 nurses. Each new patient receives a monitor that permits him or her to upload blood glucose readings which are graphed to detect trends. If a trend is found, the nurse can e-mail the patient and suggest changes in medications. Most patients see their physician every three months.

Source: A doctor, 700 patients and the net: Inventing the virtual house call. *The Wall Street Journal.* January 17, 2000: B1.

Questions for Discussion

1. Does an online service such as this one help to motivate patients with chronic diseases like diabetes to take better care of themselves? How?
2. Might an online service for patients with chronic disease cause delays in patients seeing their regular doctors? How could this be prevented?
3. Can physicians and nurses accurately assess a patient's condition and offer treatment suggestions on the basis of online data that they receive without actually seeing and examining the patient in person? If not, what would make it possible?

Case 2.12: Dial-Up Medicine

Description

John G, a 73-year-old man in La Selva Beach, Calif., was hospitalized with congestive heart failure, a disease that deprives the body of oxygen-rich blood. At times, Mr G is too tired to stand up and finds it difficult to breathe. After discharge, he lives at home, where his condition is monitored by a Web service offered by a company called LifeMasters. Each morning, Mr G weighs himself and takes his pulse and blood pressure. Next he uses his computer to log on to the LifeMaster system, where he adds his vital signs to his daily chart. Once a month, his physician is provided with a printout of his chart for review. In addition, a nurse calls Mr G every Monday to review his chart, to talk to him about his diet and exercise, and to answer questions he may have.

LifeMasters has contracts with health maintenance organization (HMOs) across the country. It also offers its monitoring program directly to the public for $200 per month through an Internet health Web site, healthnetwork.com.

Source: Fischman. J. A logon a day keeps the doctor away. *US News & World Report.* October 25, 1999: 65.

Questions for Discussion

1. What is the optimal relationship between remote-presence care and hands-on care?

2. Are there certain medical conditions that are inappropriate for online care? What makes them inappropriate: risks to the patient, family members, public health, or other factors?
3. Will some patients be intimidated or confused by the technology and unable to use it properly?

Case 2.13: Cybercare

Description

After a lifetime of good health, Mrs H was diagnosed with cancer. A 60 years-old widow, the Milwaukee woman was alone in her home. After a radical mastectomy and on chemotherapy, she found herself frightened, confused, and unsure what to expect or even what questions to ask.

Support came to Mrs H via computer. She joined the University of Wisconsin Comprehensive Health Enhancement Support System (CHESS) Network. CHESS is a computer-based support system designed to provide people facing major illnesses with information, social support, decision support, and referrals. The system enabled Mrs H to contact via the Internet fellow patients who had undergone treatment for breast cancer, and to access information from the National Library of Medicine. The program also provided her with software that helped her to choose among different treatment options.

Source: Cowley G. The rise of cyberdoc. *Newsweek*. September 26, 1994: 54–55.

Questions for Discussion

1. Is there a danger that patients will replace standard medical care with online health services such as CHESS? What might be the consequences?
2. If a consumer with a medical problem is misdiagnosed as a result of using an online health service, can the sponsor of the Web site be held liable? What should be the remedy?
3. Are HMOs and managed care groups likely to substitute less costly online services for traditional medical services? Is this unethical? Why? Why not?
4. Are some patients more comfortable discussing certain problems such as sexual dysfunction and alcohol and drug addiction online than in person with a health care provider? Is this acceptable? Why or why not?
5. Are computer programs that discuss alternative treatment options with patients less biased than physicians? Is this a good thing? Why or why not?

Case 2.14: Avoiding Drug-Drug Interactions

Description

Ms S suffered from bladder infections. When her physician prescribed cotrimoxazole, a combination of two antibiotic drugs, sulfamethoxazole and trimethoprim, Ms S used the Internet to look up the drug's profile in the

Consumer Reports Drug Database. She discovered that cotrimoxazole (Bactrim, Septra, etc.) could interact with the blood thinner Coumadin, which she was already taking to prevent blood clots in her legs. The combination of the 2 drugs could have resulted in a fatal hemorrhage. She also found that cotrimoxazole had a number of potential side effects, including rashes, itching, and severe sunburn after a short exposure to sunlight.

Ms S sent an e-mail message to her physician and included copies of her findings. She also suggested that a drug she found in her search, amoxicillin, might be a better alternative. Her doctor agreed and changed the prescription.

Source: Ferguson T. *Health online*. Reading, Mass: Addison-Wesley Publishing, 1996: xvii.

Questions for Discussion

1. Ms S might be said to have reduced her risks and improved her health care, but will patients who use the Internet to propose alternative treatments threaten their providers' self esteem and erode the professional-patient relationships? Why or why not?

2. Ms S might have made a mistake, causing herself unnecessary worry and, perhaps, wasting her physician's time. How many patients are able to understand and evaluate information about alternative drugs and treatments without the assistance of a trained professional? Should this be encouraged? Why or why not?

Case 2.15: An Interactive Support Group for ALS

Description

Tracy and James W. travel to their parents' home once a week to assist their mother in the care of their father, who has ALS (Lou Gehrig's disease). When they began caring for him, they found it difficult to obtain up-to-date information on the disease and their dad's condition.

After they discovered the ALS Digest, they began receiving online a weekly interactive support-group newsletter. Participation in the support group also permits them to exchange experiences with others who have family members with ALS. The support group helped them find a computer-assisted device that allows their father to communicate even though his vocal cords are paralyzed. They also learned of a clinical trial for a potential anti-ALS drug and enrolled their father.

Source: Ferguson T. *Health online*. Reading, Mass: Addison-Wesley Publishing; 1996: xvii–xviii.

Questions for Discussion

1. Online support groups and chat rooms are hailed as sources of information and succor for patients and their family members. Identify and evalu-

ate the potential "downside" of such services; for instance, a false sense of security (or alarm) or erroneous "information."
2. What kinds of notices or disclosures are adequate to protect people from these risks?
3. Are such notices or disclosures patronizing or paternalistic? Why, or why not?

Case 2.16: Information and Support on the Internet

Description

ZS, at age 11, was diagnosed with a severe and debilitating form of juvenile arthritis. He weighed only 59 lbs and, at times, could not walk. Since his family had only recently moved to a small town, ZS had no close friends.

While using the Internet, he and his family found the Web site for the Arthritis Society of Canada. This Web site provided information about the disease; demonstrations of exercises that would help ZS to stay active using cartoon figures; papers by top researchers in the field; symptoms of the disease; and other information on how to improve the quality of life. ZS and his family used the Web site to work with his physician to gain a better understanding and to take charge of his arthritis.

The family also found the Open Forum, a Web site that allows patients suffering from arthritis to post messages to one another. When ZS was facing surgery he wote that he was frightened, and arthritis sufferers from all over North America responded to offer him reassurance. ZS said "I found more support from this one Web site than anywhere else."
Source: McClelland S. Users beware: 'quack' sites lurk among many good Internet health links. *Maclean's*. June 21, 1999, 58–59.

Question for Discussion

Discussions about ethics are often cautionary. Sometimes, however, we can identify positive duties that humans have to each other. Concurrent with effects to protect consumers from inappropriate or inadequate online "support," might there also be a duty to encourage or even fund those online groups with positive track records?

Case 2.17: Fear and Communication

Description

The wife of a man in Ohio who had heart disease was terrified and reluctant to talk to others about her anxiety. Instead she turned to the Web for support by contacting MendedHearts Inc, a program affiliated with the American Heart Association, which operates the Web site, MendedHearts.org.

The Web site matched her with a support counselor who had undergone quadruple-bypass heart surgery himself. Over an 18-month period, the woman

carried on an e-mail correspondence with her counselor who provided information about bypass surgery and advice on recovery when the Ohio woman
and her husband needed it.

Online support programs are growing rapidly. One study has shown that
breast cancer patients had a better prognosis if they participated in a support group. Also some patients find it difficult to disclose their fears or to
ask some questions (such as if they can resume sex after bypass surgery) in
person.

Source: A guide for patients who turn to the Web for solace and support. *The Wall Street
Journal*, September 17, 1999: B1.

Questions for Discussion

1. Is the Web an appropriate surrogate or replacement for interlocutors in
   cases of shyness, fear, and/or embarrassment? What special precautions
   are needed to protect people who are thus especially vulnerable?
2. Counseling of any sort entails and creates special relationships between
   experts and clients. What rules and standards are needed to govern online
   counseling relationships?
3. Is there a danger that participants in online support groups will be subjected to angry outbursts (flaming), commercials for products and services (spam), and testimonials about miracle or unproven cures? Can this
   be prevented? How?

Sale of Body Parts

Case 2.18: Internet Bids for a Kidney

Description

A fully functional kidney was advertised for sale on eBay, an online
auction house. The ad stated "You can choose either kidney. Buyer pays all
transplant and medical costs. . . . Of course only one for sale, as I need the
other one to live. Serious bids only." Initial bidding began at $25,000; how-
ever, potential customers bid the price up to $5.7 million before bidding was
stopped.

In the United States, federal law prohibits buying or selling of human
organs. However, the Internet enables transactions of any kind, even those
that violate ethical codes and laws. While eBay has ruled against the auction-
ing of illegal products on the Web site, it does not screen items and products
offered on the site in advance. Instead, the site relies on customers to inform
managers of violations

Source: BBC News. Kidney sale on Web halted. September 3, 1999. Available at http://
news.bbc.co.uk/hi/english/sci/tech/newsid_437000/437504.stm. Accessed Oct. 3, 2001.

Questions for Discussion

1. Should people offer body parts for sale over the Internet? What is the

ethical difference between offering a kidney for sale in cyberspace and making the same offer in a traditional auction house?

2. What is to prevent Web sites based in other countries where buying and selling organs is not against the law from advertising over the Internet?

3. Should health professionals provide services to patients who obtain organs over the Internet? Why, or why not?

Case 2.19: Sperm Bank Advertisements on the Internet

Description

A sperm bank in California went online in 1997. Women looking for a sperm donor can learn the donor's ethnicity, blood type, hair color, eye color, complexion, height, weight, and education. They also can view additional information on the donor for an additional fee.

Source: Sperm bank ads on Internet. *Comput Med*, 1998;27:1.

Questions for Discussion

See Case 2.20

Case 2.20: Models' Eggs for Sale

Description

A Web site auctioned the eggs of beautiful women to the highest bidders. Opening bids were $15,000.

The Web site was designed to appeal to individuals who wish to reproduce with those they consider to be genetically superior to produce babies with social and other advantages. The donors are said to be models and actresses. Potential purchasers could view pictures of the models on the Web, and for a fee, obtain personal information about them. The sponsor of the Web site indicated that he also planned to auction sperm from handsome young men.

The bid price does not include medical costs. However, the site does list specialists who might be willing to perform the reproductive procedures after an agreement between the buyer and seller is reached.

Source: C. Goldberg, On Web models auction their eggs to bidders for beautiful children, *The New York Times*, October 23, 1999, A10.

Questions for Discussion

1. Does use of the Web to sell human reproductive material promote what has been called "naïve genetic determinism," or among other things, the false notion that children will just be like their parents especially in socially desirable ways? Why, or why not?

2. How does the sale of eggs on the Web exacerbate the reproductive exploitation of women?

3. Why do you think cases like this are more plentiful now than in pre-web

days? Is it merely the parallel evolution of two technologies (the Web and assisted reproduction), or might it be some property of cyberspace that serves as the prime stimulus?

Suicide on the Web

Case 2.21: Suicide Web Site

Description

A Dutch Web site has posted a guide to suicide methods. Step-by-step instructions guide the reader through various methods, slashing your wrists, taking sleeping pills, jumping off of buildings, and inhaling carbon monoxide, among others. For each method, the site provides the success rate along with advantages and disadvantages.

The Web site includes a disclaimer that the information is not based on expert knowledge and that the site's intention is not to encourage anyone to kill him or herself.

Source: Deutsch A. Suicide Web site sparks controversy in Netherlands. The Associated Press, January 31, 2000. Available at: http://www.chl.ca/TechNews0001/31_suicide.html. Accessed Sept. 30, 2001.

Questions for Discussion

1. Should the use of the Internet to post suicide instructions be regarded as a form of protected free speech, or should this use of the Internet be prohibited? Why, or why not? How is the online availability of suicide information different than availability in other media?
2. How could the site pose a danger to minors? How should this be weighed in considerations raised by Question 1?
3. Does the site have the potential to discourage clinically depressed persons from seeking professional help? If so, what steps, if any, would be appropriate to protect such persons?

Online Medical Records

Case 2.22: Secure Medical Records on the Internet

Description

Medtegrity Inc provides software that lets health care providers and insurers post medical records on the World Wide Web. The technology gives patients limited access to their records. A pilot project will include the medical supplier Baxter International Inc, the drug firm Merck, and insurers Aetna US Healthcare and UnitedHealth Group.

Medtegrity provides public key encryption software so that when two keys are matched, patients can find test results or check on drug prescriptions over the Internet. Medtegrity also can produce a log showing when personal medical records have been viewed.

Source: Snider K. Firm will put secure medial files on 'Net'. *The Tennessean*. Available at: http://www.tennessean.com/sii/00/05/02/medweb02.shtml. Accessed September 30, 2001.

Questions and Discussion

Note that Cases 2.22–2.26 are closely related, and questions accompanying each of them might profitably be raised for other cases in the group.

1. Surveys by the Amercian Medical Association indicate that less than 40% of practicing physicians use the Internet. Will the majority of physicians use online medical records? If there is evidence that such records can improve care, is it blameworthy for physicians not to use them?
2. Will some patients refuse to have their sensitive health data places online? With what consequences?

Case 2.23: Cyber File Cabinet

Description

Dr S became the first client of his start-up company, 4HealthyLife.com. Dr S is the founder of an Internet company that provides online storage of medical data for individuals. The firm's goal is to eliminate the need for providers to have to search for and request important medical data on a patient from a fragmented health care system.

One weekend, Dr S was admitted to an emergency room in Illinois, suffering from severe abdominal pains. He had undergone spinal surgery at the same hospital less than a week earlier. Nevertheless, the record of that surgery was not available to the emergency room physician because the medical records department was closed for the weekend. Rather than having to take a new medical history and order new laboratory tests, a nurse logged onto the 4HealthyLife.com Web site, using Dr S's identification and password, and retrieved his medical record.

Source: Raymond J. The cyber file cabinet. *American Demographics*. July 2000, 38–40.

Questions for Discussion

1. Who should be responsible for the creation and updating of online medical records? Patients? Health professionals? Hospitals or other institutions?
2. Will health care providers accept an online medical record as legitimate?

Case 2.24: Medical Records on the Internet

Description

Ms C, a 43-year-old living in Vancouver, BC, has epilepsy and takes medication regularly to prevent seizures. She joined PeronalMD.com, an online service that permits her to put her medial record on line. Her medications change from time to time, and she wants emergency room staff to have immediate access to this information if she becomes ill or is injured while traveling.

PersonalMD.com offers consumers the option of storing and accessing their own medical records using the Internet. Consumers log on to the Web site, then select a log-in name and a personal identification number (PIN). Next, they enter the information they wish to store in their medical records: demographic data, insurance coverage, emergency telephone numbers, blood type, current medications, drug allergies, and so on. Consumers can also fax documents (such as electrocardiograms and living wills) to the company to be included in their record.

Subscribers to the free service receive a wallet-size plastic identification card. The card contains emergency telephone numbers, the Personal MD.com Web address, their log-in and their PIN, allowing healthcare providers to access the patient's record in an emergency. The consumer is responsible for keeping the record current.

Source: Chase M. Patients' next choice: whether to keep files stored on the Internet. *The Wall Street Journal.* August 16, 1999: B1.

Questions and Discussion

1. It is well established that people should have access to their credit records in order to correct errors. Such an approach has not yet been worked out fully for online health data. Should patients be given unlimited access to their medical records, including provider notes?
2. What steps are needed to evaluate requests to alter online records?

Case 2.25: Medical Records Online

Description

Two New York physicians have launched an online health registry. An individual can create a personal medical record on the Web. The registry contains a complete medical history as well as digitized electrocardiographic, magnetic resonance, and x-ray images. Professionals can access the record during a routine visit or in an emergency.

The registry charges an annual fee of $100 to maintain the record on the Web site. Patients can update their records whenever necessary. These records are encrypted and stored on a secure server, and only physicians with the patient's password can access the information. The company also sells prescription drugs and medical supplies on its Web site, and offers a patient referral service, as well as an interactive service called Ask-A-Physician.

Source: Spring T. Put your medical records online. *PC World*, December 3, 1998. Available at http://www.pcworld.com/news/article.asp?aid=8928. Accessed September 30, 2001.

Questions for Discussion

1. Is an online medical record vulnerable to hackers? How can this be reduced?

2. Is there a danger that patients may omit essential medical information such as their HIV-status and/mental health problems? What is the remedy?
3. How can services that maintain online medical records ensure that the patient's record is up-do-date?
4. What is to prevent online services that store medical records from selling this information to companies who will use it for direct marketing of their products and services?

Case 2.26: Doctor Posts Patient Files on the Net

Description

Dr M, a physician who works at a community health clinic in West Virginia, is putting his patients' records on the Web. He and 3 other physicians at the clinic are electronically recording their notes from patient visits. They plan to add a problem list, a summary of conditions and allergies, hospital and medication information, laboratory results, and digital facsimiles of X-rays. So far, the system links 2 clinics and 2 affiliated hospitals.

Using the system, physicians can pull up their patients' charts and test results in the clinic or at the emergency room of the hospital, and download ultrasound images from the radiology department of a hospital located 25 miles away. A video camera and a microphone, mounted on each computer terminal, will facilitate teleconferencing, allowing doctors to consult with other physicians at multiple locations. The system will support health care for 10,000 people in a 300 square-mile rural region.

Source: Knecht GB. Click? Doctor to post patient files on net. *The Wall Street Journal*. February 20, 1996: B1,B5.

Questions for Discussion

1. Will patients have enough confidence in the security of their medical information to agree to have their records placed on a Web site? Why, or why not? How could this be improved?
2. If enough patients refuse to consent to having their records online, will this lack of data severely limit the usefulness of the system? What are the consequences?

Pharmaceutical Products

Case 2.27: Drug Bazaar

Description

A physician put Larry B's son on Accutane, a powerful prescription drug for acne. Mr B believed that it was unnecessary to pay for a monthly office visit to the doctor to obtain a prescription for a refill and for blood tests to monitor for liver damage. "Now you don't want to kill the kid; he's your kid," Mr B said. "But why pay $75 each month, when the pills only cost a few bucks?"

But Mr B is able to obtain the drug over the Internet. He now owns and operates DrugQuest, a Web site that is advertised as "the Internet doorway to money-saving medication connections around the world." The site has lists of pharmacies that are eager and willing to send customers Accutane, Prozac, thalidomide, and nearly any other prescription drug.

Source: Fischman J. Drug bazaar. *US News & World Report.* June 21, 1999: 58–62.

Questions for Discussion

1. Does the sale of prescription drugs over the Internet undermine licensing and regulation of the sale of drugs by the FDA? Why, or why not?
2. Are consumers who purchase drugs over the Internet placing themselves at risk? Should they be allowed to assume such risks? Why, or why not?
3. Is there a potential problem of quality control when consumers purchase drugs over the Internet? How could it be remedied?

Case 2.28: The Internet Diet Doctor

Description

Dr H, a Harvard-trained physician in Baltimore, Md, developed a Web site that offered diet drugs such as phentemine and fenfluramine. Hundreds of patients ordered drugs from his Web site. According to Dr H's protocol, a patient should take fenfluramine if he or she is experiencing symptoms of low serotonin such as hostility, panic attacks, and psoriasis and other autoimmune disorders, and also if he or she is experiencing cravings for carbohydrates, sex, or cigarettes.

Since Maryland law requires a physician to examine patients before prescribing medications, Dr H was indicted on 34 counts of illegally prescribing medicine. Dr H defended his practice on the grounds that the drugs that he prescribed weren't addictive, and that a physician examination was unnecessary in the cases that he treated. However, one patient that he treated died from "drug intoxication", and another committed suicide. Dr H denies any wrongdoing in these two patient deaths.

Source: James M. Former diet doctor arrested in W. Va. Baltimore Sun, July 8, 1999:1B.

Questions for Discussion

1. Is it necessary that a physician actually see the patient in person before prescribing a drug over the Internet? If yes, for all drugs? Why?
2. Should physicians who prescribe medications over the Internet without seeing a patient be held liable if the patient is harmed by the medication?
3. If a physician comes to know a patient in person, is it appropriate later to prescribe over the Internet? Why?

Case 2.29: Dispensing via the Internet

Description

Direct Response Marketing is a British-based a Web site that sold $2.5 million dollars worth of drugs, mostly Viagra, in the first 18 months of operation. Mr O, the operator of the Web site, is neither a physician nor a licensed pharmacist. When consumers log on to the site and request Viagra, they are asked to complete a questionnaire that asks if they have high blood pressure or heart problems and what other medications they are taking. A physician who reviews the questionnaire has stated that he turns down about 2% of the requests for medical reasons.

Approved requests for the drug are sent to a pharmacy, which is located near the offices of Direct Response Marketing. Mr O picks up the filled prescriptions and mails them to consumers in a plain brown envelope.

Direct Response Marketing is also offering two other drugs, Xenical, a diet drug, and Propecia, prescribed to induce hair growth. Lifestyle drugs such as these are ideal for sale over the Internet, because many consumers are embarrassed to purchase them in person. A number of drug manufacturers maintain that it is perfectly legal to supply drugs to pharmacies that sell drugs over the Internet. Furthermore, they say, any attempt on their part to influence the selling practices of licensed wholesalers would be illegal under restraint-of-trade laws.

Source: Cohen LP. Drug maker protests dispensing via Internet but practice flourishes. *The Wall Street Journal.* Nov. 29, 1999: A1,A16.

Questions and Discussion

Note that Cases 2.29–2.35 are closely related, and questions accompanying each of them might profitably be raised for other cases in the group.

1. Is a person or company that makes prescription drugs like Viagra available over the Internet recklessly endangering patients?
2. Why not let consumers decide whether they should use "lifesyle drugs?"

Case 2.30: Dying for Sex

Description

Shortly after the FDA approved Viagra, Mike H awakened at 5:00 AM. Mr H, aged 65 years, had been suffering from impotence. Intending to surprise his wife, he took a Viagra pill and went back to sleep. Later, just as he and his wife finished having sex, Mr H passed out. Two days later, he died from cardiac arrest and brain damage. His wife states that she has no doubts that Viagra contributed to her husband's death.

Immediately after FDA approval, Web sites began advertising the drug Viagra. Some sites stated that no prescription was needed, and that the drug would be shipped within 48 hours after receipt of an order. Consumers generally pay a small consulting fee, fill out an online questionnaire, indicate that they have been informed of the potential risks of using Viagra, and provide a credit card number. The consultation costs about $75, and the charge is about $10 per pill. As many as 300 pills can be purchased at one time from some Web sites.

Source: Brownlee S, Shultz S. Dying for sex: the FDA approved Viagra quickly—perhaps too quickly. *US News and World Report.* January 11, 1999: 62–66.

Questions for Discussion

1. Is there any way to prevent men with serious cardiovascular disease from purchasing Viagra from online pharmacies without first seeing their physicians?
2. Is there any way to verify the information that consumers provide in online health questionnaires when they purchase prescription drugs online?
3. Is it appropriate for a physician to approve a prescription for Viagra without examining the patient first?

Case 2.31: Internet Prescriptions

Description

A growing number of Web sites provide prescription drugs to patients. Especially popular are Viagra for impotence, Propecia for baldness, and Zenical for weight loss. The patient can fill out a questionnaire online, which is reviewed by a doctor. Then, after providing a credit card number the patient receives the medication without ever seeing a doctor.

Online pharmacies argue that they are no different from mail order pharmacies. Patients who may be too embarrassed to seek treatment in person can obtain help with the treatment of the medical problem online. Professional societies are increasingly warning physicians about the risks of online prescribing.

Source: Baldwin G. AMA warns doctors on dangers of Web pill pushing. *American Medical News,* July 19, 1999. Available at http://www.ama-assn.org/sci-pubs/amnews/ pick_99/biza0719.htm. Accessed October 1, 2001.

Questions for Discussion

1. What kind of risk is there to the patient when a doctor writes a prescription that is not based on a complete medical history or physical examination?
2. Since the doctor has no way to verify the accuracy of the information that the patient provides, is there a danger that the prescription may be inappropriate? How can this be prevented?
3. If the patient is harmed by the online prescription, is the doctor who approved the prescription, the Web site that sells the drug, or the company that distributes the drug liable? Why?

4. Should drug companies cut off access to drugs by Web sites that sell drugs without a doctor's examination and prescription?

Case 2.32: Online Importing of Drugs to the United States

Description

A Web site called Drug Quest advertises that it can help Americans "buy almost any drug without a prescription." The Web site claims the drugs are offered at "drastically discounted rates" and that 95 percent of their shipments arrive unmolested by US Customs officials. The operators of the Web site state, "We guarantee to send your order discreetly packed without any reference to the contents on the outside of the packet." They go on to warn that medicines sent by post are more likely to be intercepted by the authorities, but if the order is stolen or confiscated, they promise to refund the purchase price of the order.

A customs inspector at Kennedy International Airport in New York said that customers could buy just about anything they want over the Internet and have it shipped to their home. Inspectors can't examine every parcel that comes into the United States, and some packages carry false custom declarations. For example, unapproved drugs such as GBL (gamma butyrolactone), which are advertised on the Web to build muscle, enhance sex, reduce stress, and induce sleep, are imported into the United States labeled as wood softener or bubble bath on the customs declaration. If false custom declarations are discovered, officials seize the drugs and send a warning to the purchaser. If the shipment includes a large quantity of drugs with a high potential for resale, customs officials may deliver the drugs and arrest the buyer.

Source: Pear R. Online sales spur illegal importing of medicine to the US. *The New York Times.* January 10, 2000: A1,A12.

Questions and Discussion

1. With online pharmacies operating in many countries, is there any way that any nations's customs officials can effectively prevent the illegal sale of drugs to their citizens?
2. Is there any way that consumers can determine the quality and purity of drugs purchased over the Internet?

Case 2.33: Foreign Sales of Drugs on the Internet

Description

Electronic commerce in prescription drugs has grown rapidly. Many of the Internet pharmacies are set up in foreign countries to avoid US laws, since 80% of their customers are in the United States. The average order is about $200.

One Internet pharmacy was Vitality Health Products, in Bangkok, Thailand. Their Web site promoted prescription-free pharmaceuticals by e-mail at incredibly low prices. Some of the popular medications they supplied were Minoxidil and Propecia for hair loss; Viagra and Yohimbine, for erectile dysfunction; Retin-A and alpha hydroxy acid creams for aging skin; and testosterone and Premarin for hormone replacement. The Web site advertised that it would send the drugs in packages without any information concerning their contents on the outside of the package and without a return address. The site also told customers what to do if custom officials seized the order.

US Customs Service agents and Thai authorities raided suppliers of online pharmacies and arrested 22 people in Thailand and accused them of violating Thailand's drug and export laws. Computers and drug products were seized. In the United States, 6 people who bought drugs from the online drug company were arrested.

Source: Pear R. US and Thai officials attack sales of medicines on Internet. *The New York Times*. March 21, 2000: A1,A18.

Questions for Discussion

1. Should US or European citizens who purchase drugs illegally over the Internet be subject to criminal penalties? Why, or why not?
2. Should US Web site operators that advertise prescription medications be required to obtain state licenses? What difference would that make?
3. Are the high prices charged for drugs in the United States one of the major factors that encourages the purchasing of drugs from foreign online pharmacies? What is the remedy?
4. Should pharmacies that fill prescriptions from sites that prescribe drugs illegally be held legally liable? Why? In what way?

Case 2.34: Online Auction for Pharmaceuticals

Description

An Internet auction site, Pharmabid.com, will permit physicians and hospitals to bid for blood products, drugs, and medical supplies. While other auction sites sell health care equipment and medical supplies, Pharmabid, sponsored by the drug distributor Bergen Brunswig Corp, is the first to offer pharmaceutical supplies. The company stated that they plan to reduce variations in prices that result from local shortages of certain products like blood plasma. The Web site will initially offer about 50 products, including blood products needed to treat trauma patients, and flu vaccines.

Pharmabid uses rules by which the seller states an opening price and bidding increase progressively. Bidders, who must be licensed health care providers, may also enter a bid that indicated the maximum price they are willing to pay for the product. Winning bidders are notified by e-mail, and the products are delivered the next day.

Source: Rundle RL. Pharmaceuticals to be auctioned on Web. *The Wall Street Journal*. December 1, 1999: B6.

Questions for Discussion

1. Considering the high cost of drugs, should consumers be allowed to bid on drugs that they need for their health care? Why, or why not?
2. Do health care providers who purchase pharmaceuticals at a discount from an online auction have responsibility to pass on some of the savings to patients?

Case 2.35: Drug Database Sells Advertisements to Sponsors

Description

A Web site, ePocrates, offers free information on prescription drugs. Clinicians can download to personal digital assistants data on dosages, adverse reactions, and other information on about 1600 medications. This permits them to have access to these data anywhere in their offices or the hospital and at the patient's bedside. So far about 70,000 users have downloaded data from the site.

To continue offering the service, the company plans to contract with drug companies, managed care firms, and other sponsors that will pay to display ads on the ePocrates network. The system will also send messages to clinicans, which they can pass on to their patients, concerning alternative medications and how the patient can acquire the drug at a discount.

Source: Rundle R. Drug database for doctors sells ads to sponsors. *The Wall Street Journal*. June 19, 2000: B1.

Questions for Discussion

1. Is there a conflict of interest when a pharmaceutical company that sponsors a Web site supplies information about drugs? How could it be resolved?
2. How should physicians evaluate the accuracy and validity of information provided by a drug company's advertisements?

Mental Health Services

Case 2.36: Diagnosing Depression by Computer

Description

The National Mental Health Association in Alexandria, Va, sponsors a Web site to screen for depression, www.depressionscreening.org. Patients can answer a set of questions on a diagnostic instrument developed by Harvard University and the National Mental Illness Screening project. A score is printed out for the patient, who can then take this information to a mental health professional. The Web site also offers a list of affiliated mental health associations and caregivers, and links with the American Psychiatric Association, the American Psychological Association, the National Association of Social Workers, and the American Psychiatric Nurses Association.

Sponsors of the Web site say use of the online service isn't a substitute for a face-to-face evaluation by a licensed therapist. Rather, the questionnaire can be used as a first step toward obtaining help by determining if the patient is clinically depressed. Moreover, patients can use the questionnaire to monitor their symptoms and determine the efficacy of antidepressant drugs between visits to their therapist.

Source: Chase M. Diagnosing depression by computer may spur more to get treatment. *The Wall Street Journal.* November 19, 1999: B1.

Questions for Discussion

Note that Cases 2.36–2.38 are closely related, and questions accompanying each of them might profitably be raised for other cases in the group.

1. Is there a danger that online screening may aggravate a sense of isolation and depression in some patients? If so, what strategies or procedures are needed to reduce this adverse effect?
2. Can patients or clients receive adequate workups and evaluations online? If so, are any special precautions needed?
3. Will some patients who use the online service avoid seeing a therapist?
4. Should mental health professionals receive special training in online practice?
5. Is the information collected online concerning a patient's psychiatric condition adequately protected from privacy violations?

Case 2.37: The Cybercouch

Description

In New York, a group of 5 therapists (4 psychologists and 1 psychiatrist) formed an electronic practice called Shrink-Link to provide mental health services. Each therapist has a regular practice and specializes in an area of psychotherapy. For example, one practitioner specializes in substance abuse and couples therapy.

The service charges $20 for each contact. Patients send a question of 200 words or less over the Internet and enter a credit-card number. The Web site promises an e-mail response, consisting of 3 paragraphs of advice, within 72 hours. An example of a question is, "My 5-year-old was diagnosed with attention deficit disorder (ADD) in 1993 and has been on Ritalin ever since. She has been having trouble falling asleep for the past several months and has been moodier than usual of late. What do you think?" The gist of the response was: Some trial and error is often required before the correct dosage and timing are found, and symptoms such as sleep disturbance and moodiness often occur in the interim. Moreover, since children's rates of metabolism change, dosages often need to be adjusted. Even if the dosage is correct, the behavior irregularities could be caused by administering the drug too late in the afternoon or by a host of other factors, such as nighttime fears. These possibilities need to be ruled out one by one until the culprit is found.

Source: Hannon K. Upset? Try cybertherapy. *US News & World Report.* May 13, 1996: 81–83.

Case 2.38: Counseling on the Internet

Discussion

Concerned Counseling, www.concernedcounseling.com, is a network of therapists who use the Internet to provide counseling to individuals and families. Topics include marital conflict, eating disorders, and drug and alcohol addiction. Counselors are available for chatroom consultation 12 hours per day, seven days per week. Patients may send e-mail messages that are generally answered within 12 hours. The cost of a chatroom session is $45 for up to 30 minutes, $80 for up to 50 minutes. Each additional 15 minutes costs $25. E-mail consultation costs $30 for each response.

Consumers of the service must agree to abide by certain rules. These rules cover online conduct, copyright restrictions, limitations on liability, and warranty and indemnification provisions. The site states that counselors are licensed therapists but it does not provide details concerning their academic and professional credentials.

Source: LeBourdais E. When medicine moves to the Internet, its legal issues tag along. *Can Med Assoc J.* 1997;157:1431–1433.

Questions for Discussion

1. How can patients verify a therapist's qualifications to provide mental health services?
2. Are patients who avoid seeking help in person for psychological problems more likely to seek online services? Is this a good thing?
3. Are patients who consult an online service less likely to consult a licensed therapist in person? Why?
4. How can the patient's confidentiality be protected if the service is not certain who is reading the e-mail messages or where they are stored?
5. When therapists do counsel patients over the Internet, should copies of e-mails and transcripts of chatroom sessions be filed with the patient's medical record?

Cyberchondriasis

Case 2.39: Tingling Feet or MS?

Description

Health-related Web sites provide a new source of information for actual or potential hypochondriacs. First-person testimonials providing extensive fodder for the would-be ill are shared in chat rooms and on bulletin boards.

Combined with the vast amounts of information available on any disease or disorder, such encouragement can feed a person's anxiety. One woman who noticed a tingling in her feet surfed the Web in search of a diagnosis of her problem and became convinced that she had multiple sclerosis when she found a Web advertisement for multiple sclerosis medication.

Source: Carrns A. Cyberchondriacs get what goes around on the Internet now. *The Wall Street Journal.* October 5, 1999: A1, A6.

Questions for Discussion

1. Are potential hypochondriacs more likely to be drawn to the Internet for information concerning medical conditions they fear?

2. Does information on the Internet about exotic diseases unnecessarily aggravate anxiety more than other sources of information?

Case 2.40: Münchausen Syndrome and Medical Record Hacking

Description

A 32-year-old man seeking experimental cancer therapy in Pennsylvania showed up with medical records from a hospital in Florida. The records included surgical pathology, cytology, and radiology reports describing liver and other metastases, but these reports conflicted with the results of a physical examination. Because of the discrepancies, physicians in Pennsylvania sought confirmation from their Florida counterparts.

Florida physicians found a number of discrepancies between the copy of the surgical pathology report and the original record that indicated tampering. A search of the hospital's computer system found that the man had been evaluated in the emergency room of the Florida hospital once for flu-like symptoms and once for urinary complaints. He had not been biopsied, and there was no record that the patient had been diagnosed with cancer. The center where he had reportedly received a bone marrow transplant had no record of caring for him.

A further search of the computer system found that the medical report presented in Pennsylvania was a composite of two or more diagnostic reports for other patients. Portions of the reports, including the institutional logo and format, had been electronically scanned, recombined, edited, printed, and copied. Logs at the Florida hospital indicated that the patient had reviewed his own chart 1 week before he referred himself to the Pennsylvania center. He had probably introduced the extraneous material into his chart at that time.

Source: Hadeed V, Mies C, Wunsch CD, Trump DL. Münchausen syndrome: Electronic records and cancer (letter). *Ann Intern Med.* 1998;129:73.

Questions for Discussion

1. Some forms of computer abuse can rarely be defended against completely. To what extent should institutions take steps to prevent the unlikely but severe consequences of pathological behavior?

2. Will the existence of electronic medical records make it easier for patients to fake medical conditions either as an expression of a behavioral malady or for fraudulent or other illicit purposes? What preventative measures should be taken?

Case 2.41: Münchausen Syndrome Online

Description

A 23-year-old woman told others in an Internet eating disorder support group that she had such a disorder and was hiding from an abusive boyfriend. She also said she was in an intensive care unit. From time to time, she typed unintelligible messages, which she said were caused by periods when she was in shock. At one point, the woman feigned a stroke while online. Subsequent messages, supposedly from her mother, provided participants in the chat room with progress reports on the woman's condition. A participant in the support group, who called the hospital where the woman said she was admitted, discovered the hoax. Afterward, the same woman joined chat rooms for sexual abuse survivors and for persons with acquired immune deficiency syndrome (AIDS).

Source: Feldman MD. Münchausen Syndrome online: Tales of suffering deceive Internet support groups. *Comput Med.* 1998;27:3,5.

Questions for Discussion

1. Is there any way to identify persons with fictitious disorders who join Internet support groups in order to help them? Would this violate others' privacy or confidentiality?
2. Do Web sites that provide information about diseases indirectly or tacitly encourage or stimulate hypochondriasis or Münchausen-like behavior in certain vulnerable people?

Note: Parts of this chapter are adapted from Anderson JG. The business of cyberhealthcare. MD Computing. *1999:16:23–25, with permission.*

References

1. Goldsmith J. How will the Internet change our health system? *Health Affairs.* 2000;19:148–156.
2. Dembeck C. Online healthcare expected to reach $370 billion by 2004. *EcommerceTimes.com*, January 4, 2000. Available at: http://www.ecommercetimes.con/perl/printer/2128. Accessed October 4, 2001.
3. Stolberg SG. From M.D. to I.P.O., Chasing Virtual Fortunes. *The New York Times* Sect. 4, p. 3, July 4, 1999.
4. Kalb C, Branscum D. Doctors go Dot.com. *Newsweek.* August 16, 1999:65–66.
5. Naujeck, A. Consumers shopping on Web for surgical options. The Associated Press, May 12, 2000. Available at: http:/acmi.canoe.ca/Health0005/12_surgery.html. Accessed Sept. 30, 2001.

6. Fischman J. A logon a day keeps the doctor away. *US News & World Report.* October 25, 1999, 65.

7. Spring T. Put your medical records online. *PC World*, December 3, 1998. Available at: http://www.pcworld.com/news/article.asp?aid=8928. Accessed September 30, 2001.

8. Chase M. Patients' next cholice: whether to keep files stored on the Internet. *The Wall Street Journal*, August 16, 1999, B1.

9. Fischman J. Getting medicine off the Web is easy, but dangerous. *US News & World Report.* June 21, 1999, 58–62.

10. Cohen LP. Drug maker protests dispensing via Internet but practice flourishes. *The Wall Street Journal.* November 29, 1999: A1, A16.

11. LeBourdais E. When medicine moves to the Internet, its legal issues tag along. *Can Med Assoc J.* 1997;157:1431–1433.

12. McClung HJ, Murray RD, Heitlinger LA. The Internet as a source for current patient information. *Pediatrics.* 1998:101:E2. Available at: http://www.Pediatrics.org/egs/content/full/101/6/E2. Accessed October 1, 2001.

13. Stolberg SG. Trade agency finds Web slippery with snake oil. *New York Times.* June 25, 1999: 1, A16.

14. Rubin R. Prescribing on line: The AMA calls it 'bad medicine' but others call it the future. *USA Today.* November 2, 1998: 1A, 2A.

15. Pear R. U.S. and Thai officials attack sales of medicines on Internet. *New York Times.* March 21, 2000: A1, A18.

16. Noble HB. Hailed as a Surgeon General, Koop criticized on Web ethics. *New York Times.* September 5, 1999: 1, 18.

17. Laidman J, Woods M. Sex doctor's patient files show up on the Web. *Pittsburgh Post-Gazette*, March 28, 1999. Available at http://www.post-gazette.com/headlines/19990328doclist2.asp. Accessed October 1, 2001.

18. Harrow Jr. RO. Prescription sales, privacy fears: CVS, Giant share customer records with drugs marketing firm. *The Washington Post.* February 15, 1998: A1, A1.

19. Bulkeley WM. E-mail medicine: Untested treatments, cures find stronghold on online services. *The Wall Street Journal.* February 27, 1995: A1, A7.

Further Readings

Anderson JG. Clinical information systems. In: Kent A, ed. *Encyclopedia of library and information science*, Vol 69. New York: Marcel Dekker; 2000: 33–53.

Anderson JG. Computer-based ambulatory information systems: Recent developments. *J Ambul Care Manage.* 2000;23:53–63.

Anderson JG, Jay SJ, eds. *Use and impact of computers in clinical medicine.* New York: Springer-Verlag; 1987.

Ball MJ, Collen MF, eds. *Aspects of the computer-based patient record.* New York: Springer-Verlag; 1992.

Ball MJ, Douglas JV, O'Desky RI, Albright JW. *Healthcare information management systems: a practical guide.* New York: Springer-Verlag; 1991.

Berner ES, ed. *Clinical decision support systems: theory and practice.* New York: Springer-Verlag; 1999.

Bleich HL, Slack WV. Clinical computing. *MD Computing.* 1989;6:133–135.

Boberg EW, Gustafson DH, Hawkins RP, et al. CHESS: The comprehensive health enhancement support system. In: Brennan PF, Schneider E, eds. *Information networks for community health.* New York: Springer; 1997:171–188.

Brennan PF, Ripich S, Moore S. The use of home-based computers to support persons living with AIDS/ARC. *J Commun Health Nurs.* 1991;8:3–14.

Brennan PF, Schneider SJ, Jornquist E, eds. *Information networks for community health.* New York: Springer; 1997.

Brody JE. The health hazards of point-and-click medicine. *The New York Times.* August 31, 1999:D1.

Chase M. A guide for patients who turn to the web for solace and support. *The Wall Street Journal.* September 17, 1999: B1.

Collen MF. *A History of medical informatics in the United States, 1950 to 1990.* Bethesda, Md: American Medical Informatics Association; 1995.

Cowley G. The rise of cyberdoc. *Newsweek.* September 26, 1994:54–55.

Dick RS, Steen EB, Detmer DE. *The computer-based patient record.* Rev ed. Washington, DC: National Academy Press; 1997.

Eder L, ed. *Managing healthcare information swith web-enabled Technologies.* Hershey, Pa: Idea Group Publishing; 2000.

Engstrom PM, Brown MS. Electronic house calls: New rules, new roles as healers swarm the Net. *Med Net.* 1996;2:1–6.

Ferguson T. *Health online: how to find health information, support groups, and self-help communities in cyberspace.* Reading, Mass: Addison-Wesley; 1996.

Fischman J. Drug bazaar. *US News & World Report.* June 21, 1999:58–62.

Fitzgibbons S, Lee R. *The Health.net industry: the convergence of healthcare and the internet.* San Francisco: Hambrecht and Quist; 1999.

Freeman S. Michigan tells net pharmacies it plans suits. *The Wall Street Journal.* December 16, 1999:B2.

Gallo AC, Lee VJ. *Health care information technology: keeping health care wired. research report.* Baltimore, Md: Alex Brown; 1998.

Glowniak JV. Medical resources on the internet. *Ann Intern Med.* 1995;123:123–131.

Glowniak JV, Bushway MK. Computer networks as a medical resource: accessing and using the internet. *JAMA.* 1994;271:1934–1939.

Golden PA, Beauclair R, Sussman L. Factors affecting electronic mail use. *Comput Hum Behav.* 1992;8:297–311.

Goldsmith J. How will the Internet change our health system? *Health Affairs.* 2000;19:148–156.

Goldwein JW. Internet-based medical information: Time to take charge. *Ann Intern Med.* 1995;123:152–153.

Goodman KW. Bioethics and health informatics: An introduction. In: Goodman KW ed. *Ethics, computing and medicine: informatics and the transformation of health care.* Cambridge, UK: Cambridge University Press; 1998:1–31.

Gould RL. The use of computers in therapy. In: Trabin T, Freeman MA, eds. *The computerization of behavioral healthcare: how to enhance clinical practice, management, and communications.* San Francisco: Jossey-Bass; 1996.

Greenes RA, Shortliffe EH. Medical informatics: an emerging academic discipline and institutional priority. *JAMA.* 1990;263:1114–1120.

Hannon K. Upset? Try cybertherapy. *US News & World Report.* May 13, 1999:81, 83.

Health information revolution: Preparing for the information war. *Health Affairs.* 1998;17:9–60.

Hoffman M, Hock C, Muller-Spahn F. Computer-based cognitive training in Alzheimer's disease patients. *Ann NY Acad Sci.* 1996;777:249–254.

Johannes L. Competing online, drugstore chains virtually undersell themselves. *The Wall Street Journal.* January 10, 2000:B1.

Kasper JF, Mulley AG, Wennberg JE. Developing shared decision-making programs to improve the quality of health care. *Quality Review Bulletin.* June 1992:183–190.

Kassirer JP. A report card on computer-assisted diagnosis—the grade: C. *N Engl J Med.* 1994;330:1824–1825.

Kassirer JP. The next transformation in the delivery of health care. *N Engl J Med.* 1995;332:52–53.

Kiel JM. yourpractice.com: Making the leap to the internet. *MD Comput.* September/October 1999:27–29.

Kiel JM, Cherry JC. Positive outcomes lower costs: Using net-based IT to manage care. *MD Comput.* March/April 2000:27–28.

Kleinke JD. *Bleeding edge: the business of health care in the next century.* Gaithersburg, Md: Aspen; 1998.

Kleinke JD. Release 0.0: Clinical information technology in the real world. *Health Affairs.* November/December 1998:23–38.

Legler JD. Computers and the physician-patient relationship: What do we know. In Miller RA ed. *Proceedings of the fourteenth annual symposium on computer applications in medical care.* New York: IEEE Computer Society, 1990:289–292.

Lindberg DAB. *The growth of medical information systems in the United States.* Lexington, Mass: Lexington Books; 1979.

Lindberg DA, Humphreys BL. Computers in medicine. *JAMA.* 1995;273:1667–1668.

Locke S, Kowaloff HB, Hoff RG, et al. Computer-based interview for screening blood donors for risk of HIV transmission. *JAMA.* 1992;268:1301–1305.

Lord M. Medicine comes to the tube. *US News & World Report.* August 9, 1999:62.

Lueck S. More authority over online drug sellers is sought by state and FDA officials. *Wall Street Journal.* March 22, 2000:B6.

McConnell J. Medicine on the superhighway. *Lancet.* 1993;342:1313–1314.

McDonald CJ, et al. Canopy computing using the web in clinical practice. *JAMA.* October 21, 1998:1325–1329.

McKinney WP, Wagner JM, Bunton MS, Kirk LM. A guide to mosaic and the world wide web for physicians. *MD Computing.* 1995;12:109–114,141.

McTravish FM, Gustafson DH, Owens BH, et al. CHESS (comprehensive health enhancement support system), and interactive computer system for women with breast cancer piloted with an underserved population. *J Ambulatory Care Manage.* 1995;18:35–41.

Millenson ML. *Demanding medical excellence: doctors and accountability in the information age.* Chicago: University of Chicago; 1997.

Miller RA, Schaffner KF, Meisel A. Ethical and legal issues related to the use of computer programs in clinical medicine. *Ann Intern Med.* 1985;102:529–536.

Millstein SG, Irwin CE Jr. Acceptability of computer-acquired sexual histories in adolescent girls. *J Pediatr.* 1983;103:815–819.

Morrissey J. Internet company rates hospitals. *Modern Healthcare.* August 16, 1999:24.

Newman LA. Health plans compete in cyber-marketplaces. *Executive Solutions Healthcare Manage.* October 1999:5–6.

Nicoll NH. Grateful med: an easy to use information tool. *Reflections*. 1992;18:40–41.

Pannen M. Guide to the internet: The world wide web. *BMJ*. 1995;311:1552–1556.

Patterson TL, Shaw WS, Masys DR. Improving health through computer self-help programs: Theory and practice. In: Brennan PF, Schneider E, eds. *Information Networks for Community Health*. New York: Springer; 1997:219–246.

Pear R. Controls sought for drug sales on the Internet. *The New York Times*. December 28, 1999:A1, A19.

Ray J, Sydnor J. *Disease management: the future of managed care*. New York: Union Capital Markets; April 12, 1999.

Rind DM, Safran C, Phillips RS, et al. Effect of computer-based alerts on the treatment and outcomes of hospitalized patients. *Arch Intern Med*. 1994;154:1511–1517.

Rose JS. *Medicine and the information age*. Tampa, Fla.: ACPE Publication; 1998.

Rubin R. Industry's rapid growth, change defy regulation. *USA Today*. November 2, 1998:A1, A2.

Ruffin M. *Digital doctors*. Tampa: American College of Physician Executives; 1999.

Safran C, Herrmann F, Rind D, Kowaloff H, Bleich H, Slack WV. Computer-based support for clinical decision making. *MD Comput*. 1990;7:319–322.

Safran C, Rind DM, Davis RB, et al. Guidelines for management of HIV infection with computer-based patient's record. *Lancet*. 1995;346:341–346.

Schwartz WB, Patil RS, Szolovits P. Artificial intelligence in medicine: Where do we stand? *N Engl J Med*. 1987;316:685–688.

Scolamiero SJ. Support groups in cyberspace. *MD Comput*. 1996;14:12–17.

Selami PM, Klein MH, Greist JH, Sorrell SP, Erdman HP. Computer-administered cognitive-behavorial therapy for depression. *Am J Psychiatry*. 1990;147:51–56.

Serafini MW. Drugs on the web. *National J*. November 13, 1999:3310–3314.

Sherrid P. What's up, Dr. Koop? *US News & World Report*. September 20, 1999:51.

Shine KI. Impact of information technology on medicine. *Technology in Society*. 1996;18:117–126.

Slack WV. The computer and the doctor-patient relationship. *MD Comput*. 1989;6:320–321.

Slack WV. When the home is also the clinic. *MD Comput*. 1996;13:465–468.

Slack WV. *Cybermedicine*. San Francisco, Calif: Jossey-Bass; 1997.

Slack WV, Boro ES, Bleich HL. Barriers to clinical computing: what physicians can do. *MD Comput*. 1992;9:278–280.

Slack WV, Safran C, Bleich HL. Computerization in hospital-based delivery systems. In: Trabin T, Freeman MA, eds. *The Computerization of behavioral healthcare: how to enhance clinical practice, management, and communications*. San Francisco: Jossey-Bass; 1996.

Slack WV, Van Cura LJ, Greist JH. Computers and doctors: Use and consequences. *Comput Biomed Res*. 1970;3:521–527.

Smyth KA, Feinstein SJ, Kacerek S. The Alzheimer's disease support center: Information and support for family caregivers through computer-mediated communication. In: Brennan PF, Schneider E, eds. *Information networks for community health*. New York: Springer; 1997:189–203.

Stammer L. Physicianconnection. *Healthcare Inf*. March 2000:51–53.

Starr P. Cyberpower and freedom. *The American Prospect*. 1997;33:6–9.

Starr P. Smart technology, stunted policy: developing health information networks. *Health Affairs*. May/June 1997:91–105.

Szolovits P, Patil RS, Schwartz WB. Artificial intelligence in medical diagnosis. *Ann Intern Med*. 1988;108:80–87.

Taenzer P, Zendel I, Birdsell JM, McGregor SE, Freiwald S. Cancer, me?: health care advice goes online. In: Brennan PF, Schneider E, eds. *Information networks for community health*. New York: Springer; 1997:205–217.

Tan JKH. *Health management information systems*. Gaithersburg, Md: Aspen Publishers; 1995.

Tanouye E, Carrns A, Wingfield N. America Online and CareInsite join in an online health-services alliance. *The Wall Street Journal*. September 16, 1999:B10.

US Department of Commerce. *Falling through the net III*. Washington: US Department of Commerce; July 1999.

van Bemmel JH, Musen MA, eds. *Handbook of medical informatics*. New York: Springer; 1997.

Zelingher J. Exploring the internet. *MD Comput*. 1995;12:100–108.

3

Consumer Health Information: Let the Viewer Beware (Caveat Viewor)

"Rapid pulse, sweating, shallow breathing . . . according to the computer, you've got gallstones."

Herman® is reprinted with permission from LaughingStock, Inc., Ottawa, Canada. All rights reserved.

A rapidly growing number of individuals are seeking health-related information through the Internet [1]. The Internet provides access to a host of medical resources that can help patients make informed decisions about health care and to assume more responsibility in managing their health. For patients with life-threatening disease, Web sites and on-line support groups can provide emotional support as well as invaluable information about alternatives. On the other hand, the information available on the Internet may be bewildering, incomplete, out-of-date, or false and misleading. Although a wealth of health information is available to everyone with access to the Internet, there is no guarantee of the quality and accuracy of what they find.

Inaccurate and Out-of-Date Information

Even Web sites sponsored by trusted sources such as major medical centers and research institutions may contain inaccurate information or may contradict one another. One study investigated the accuracy of information a lay person could obtain from Internet sources concerning the treatment of childhood diarrhea [2]. A routine search with widely used search engines (Yahoo, Netscape, and Infoseek) found over 15,000 references on childhood diarrhea, an overwhelming amount of material for anyone to sort through. Of the first 300 documents retrieved, only 60 articles were published by traditional medical sources, and only 20% of these conformed to guidelines published by the American Academy of Pediatrics. A study of information concerning the treatment of a child with a fever [3] compared the information on a number of Web sites from reliable sources to the advice found in a standard medical text. Some sites contained correct information. Others recommended that aspirin be administered, placing the child at risk of Reye's syndrome. A study of the quality of surgical information available on the Internet concluded that variation in contents limits the use of such information as a reliable and safe source for health care providers patients [4].

Unverified and Fraudulent Health Claims

Web sites that promote unproven or even phony cures are proliferating. The Federal Trade Commission (FTC) has identified hundreds of Web sites promoting unproven cures for 30 ailments including AIDS, multiple sclerosis, liver disease, arthritis, and cancer [5].

The FTC has reached a legal settlement with four of the companies using the Internet to promote phony cures (One advertised a fatty acid to cure arthritis, another sold shark cartilage as a cure for cancer and AIDS, and two promoted magnets as therapy for various diseases), but the large number of sites makes it impossible for the agency to investigate and take action against most of them. Investigations are time consuming and require the marshaling of comprehensive scientific information to prove that a claim is false or advertising is deceptive. For example, several sites continue to promote and sell apricot seeds or laetrile as a cure for cancer, though the FDA has banned this supposed cure [6]. The resurgence in sales has resulted from an increased public interest in alternative treatment, especially among desperate patients. Sites such as ApricotsfromGod.com and CancerAnswer.com offer a 21-day supply of injectable laetrile and feature dozens of testimonials from patients who claim to have been cured of cancer by taking the compound. While the FDA has conducted undercover investigations and sent warning letters, it had been unable to close down the sites because substances such as laetrile can be legally sold as nutritional supplements as long as the sites do not brazenly claim that it cures cancer. In April 2000, a US District Judge issued a preliminary order barring the operator of the sites from selling laetrile and related products including apricot seeds for cancer treatment.

Potential Conflicts of Interest

An additional problem with Web sites that provide health-related information is conflict of interest. One report on health care spending indicated that, in 1998, pharmaceutical companies spent $1.3 billion to advertise new drugs directly to consumers, a 55% increase over 1997 [7]. Nowhere is this effort more apparent than on the Internet, where drug companies are investing heavily to create a dominant presence [8]. For example, the Web site of Healthtalk Interactive Inc, www.healthtalk.com, features panel discussions concerning new therapies. Grants from Glaxo, Genentech, Amgen, and Berlex Laboratories fund patient education on this site and other sites that deal with diseases for which the companies manufacture drugs. Bristol-Myers Squibb's oncology drug division supports Healthology.com's cancer awareness program. Even the *Journal of the American Medical Association's* HIV/AIDS information site is supported by Glaxo Wellcome. The blurring of commercial content and independent professional evaluation of drugs is a major concern. There are no laws that regulate the content of Web sites.

Ethics Codes and Guidelines

The health information available on the internet presents consumers with more uncertainty and variability than any other sector of the economy [9]. An important question is whether there are means to provide consumers a way to judge the credibility of health information available on Web sites. A number of efforts have begun toward the development of standards for medical and health Web sites (see Appendix). More than a dozen companies that provide health information online have formed an alliance to develop a code of conduct to assure consumers of the reliability and safety of the information they provide [10]. Guidelines have been published by health care technology groups, non-profit organizations, and journals of professional health care organizations. The Health Information Technology Institute, Mitretek Systems, has published draft criteria for assessing the quality of Internet health information [11]. Health on the Net Foundation, based in Geneva, Switzerland, an international not-for-profit organization, has published 8 standards for health and medical Web sites [12]. Two principles central to the Health on the Net code are that health advice should be provided by medically trained and qualified professionals, and that information should complement and not replace or supplant the relationship between a patient and a health care provider. Other sets of standards by which professionals and consumers may judge the credibility of health information on the Web have appeared in the *Journal of the American Medical Association* [13] and the *British Medical Journal* [14].

Rating Health Web Sites

A growing number of instruments have been proposed to rate Web sites that provide health information, frequently by producing awards, ratings of quality, and seals of approval. One study, however, found that many of these instruments do not provide explicit descriptions of the criteria used for ratings, information as to how the criteria were generated or selected, identification of the writers of the criteria, or guidance in how the criteria should be used [15].

To evaluate the content of a Web site, the evaluator or evaluating instrument needs to compare the information provided with relevant evidence-based knowledge. The dynamic nature of the Internet poses additional evaluation problems. Web sites are created and modified by many different groups. Information is presented in different formats and is linked to many other Web sites. A comprehensive evaluation must go beyond content and consider how easy it is to find the site; the impact that the information provided has on clinical processes and patient outcomes; and the cost-effectiveness of obtaining health information over the Internet compared to alternatives [13].

Assisting the Consumer

We must provide the public with tools to identify Web sites that provide reliable health information in a timely fashion. One such effort is a Web site created by Health Canada which provides general information on a large variety of health-related topics including mental health, human sexuality, and disease prevention. The site also provides links to other web sites sponsored by recognized Canadian health institutions and disease associations [16]. Another site, sponsored in the United States by the National Library of Medicine, is MEDLINEplus.gov [17]. This site provides consumers with quality information from the National Institutes of Health, the Food and Drug Administration, and other government and non-profit organizations. As difficult as these efforts will be, we must commit the necessary resources to avoid further erosion of public trust and confidence in a health care system inundated with health information.

Case Studies

Inaccurate Information

Case 3.1: Inaccurate Medical Information on the Web—Childhood Diarrhea

Description

Investigators performed a routine search on the Web for information on treating childhood diarrhea, employing widely used search engines such as Yahoo, Netscape, and Infoseek. The search found more than 15,000 sources of information. The information on each Web site was classified by its source, either traditional (e.g., academic medical center) or alternative (e.g., homeopathic, chiropractic, or naturopathic).

A random selection of 300 sources found only 60 sites that provided information from traditional medical sources. Further investigation revealed that only 20% of the 60 reputable sites contained treatment recommendations that conformed to guidelines published by the American Academy of Pediatrics. Many of the Web sites sponsored by traditional medical sources described the pathophysiology of infectious diarrhea incorrectly. Treatment recommendations from university medical centers frequently failed to distinguish the special needs of infants. In some instances, departments within the same institution such as pediatrics, family practice, infectious disease, and gastroenterology, offered conflicting information concerning the treatment of diarrhea.

Source: McClung HJ, Murray RD, Heitlinger LA. The Internet as a source for current patient information. *Pediatrics.* 1998:101:E2. Available at http://www.Pedatrics.org/egs/content/full/101/6/E2. Accessed June 4, 1998.

Questions for Discussion

1. How regularly or frequently should health Web sites review or update their information?
2. What level of institutional oversight is needed when several departments within the same institution have Web sites that provide consumer information on a common health problem?

Case 3.2: Cancer Information on the Internet

Description

A team of researchers at the University of Michigan analyzed 371 Web sites that contained information on Ewing's sarcoma, a bone cancer that afflicts children and young adults. The investigators developed an instrument to evaluate each Web page for its relevance to the topic, whether the site contained medical information, and whether the information was anecdotal or peer reviewed. In addition, all medical material contained on the site was evaluated for accuracy by 2 reviewers.

About one third of the Web sites contained references to information that did not come from peer-reviewed sources. At least 6% of these Web pages contained errors. Information on other Web pages was outdated or misleading. For example, survival rates ranging from 5% to 85% (contrast a generally accepted survival rate of 70% to 75%) were reported. Additionally, many alternative medicine and health sites gave cancer care advice, some of which was controversial.

Source: Biermann JS, Golladay GJ, Greenfield ML, Baker LH. Evaluation of cancer information on the Internet. *Cancer.* 1999;86:381–390.

Questions for Discussion

1. What duties do health professionals have to warn and educate patients and parents about misinformation on the Web? In the absence of such guidance, will parents be tempted to try unproven, inappropriate, and dangerous therapies?
2. How great is the danger that parents of children with cancer might forgo therapy if they access a Web site that reports there is a low survival rate for their child's form of cancer?
3. Are patients generally able to distinguish peer-reviewed information from less reliable data?

Case 3.3: Internet Advice on Home Management of Feverish Children

Description

An Italian research team assessed the reliability of Web-based advice to parents with feverish children. Two search engines, Yahoo! and Excite, were used to find relevant information. For each Web site located, the team recorded the type of organization sponsoring the site, the country out of which the Web site operated, and the specific information provided about managing of fever in children.

The search found 41 Web sites. Of these, 31 were sponsored by commercial interests, and 9 were sponsored by academic medical institutions. A great deal of variability existed in the information provided by the sites. Most Web sites recommended rectal temperature measurement. Thirty-one sites mentioned drug treatment while 38 recommended non-drug measures. Two sites recommended treatment of fever with aspirin but failed to warn parents about the danger of Reye's Syndrome. Only 4 of the Web sites, less than 10%, adhered closely to the main recommendations of published guidelines.

Source: Impicciatore P, Pandolfini C, Casella N, Bonati M. Reliability of health information for the public on the World Wide Web: Systematic survey of advice on managing fever in children at home. *BMJ.* 1997;314:1875–1879.

Questions for Discussion

1. How can parents effectively use the variable and sometimes contradic-tory Web-based information on managing fever in children?
2. Are parents likely to search for sources of information that conform to their preconceived ideas about treatment of fever or, indeed, any other malady?
3. Are parents who use the Web to obtain treatment information less likely to consult a physician about their child's condition?

Unverified Health Claims

Case 3.4: "Web Slippery with Snake Oil"

Description

The US Federal Trade Commission (FTC) has identified hundreds of Web sites promoting unverified cures for 30 ailments including AIDS, multiple sclerosis, liver disease, arthritis, and cancer. One advertised a fatty acid to cure arthritis; another site sold shark cartilage as a cure for cancer and AIDS; two sites promoted magnets as therapy for various diseases. In 1997, and again in 1998, the FTC and public health agencies in 25 countries scanned the Internet for fraudulent health claims. Each session identified over 400 site promoting questionable treatments. Warnings were sent to many of the sites, but 2 months later, three fourths of the sites were still operating un-changed. A year later, only 28% of the sites that were warned had dropped their claims.

The FTC stated that it does not have the resources to conduct ongoing surveillance of Web sites that make health-related claims. Investigations are time consuming and require a thorough review of the scientific literature related to the claims to prove that a Web site is promoting deceptive or phony cures. The FTC has reached legal settlements with only 4 of the companies that were identified as using the Internet to promote phony cures.

Source: Stolberg SG. Trade agency finds Web slippery with snake oil. *The New York Times.* June 25, 1999: A16.

Questions for Discussion

1. Should federal and state government try to protect consumers from un-verified health claims on Web sites? How?
2. How can consumers distinguish good from bad health information on the Web?

Case 3.5: The Growth of Unverified Health Claims

Description

More and more products of dubious value are promoted on the Internet. A search with the key words "alternative medicine" brings up hundreds of mail-order houses that supply alternative remedies. Substances available range from combinations of herbs to a menu of experimental drugs with no restrictions on who can order them from the Web site. For example, claims are made that shark cartilage inhibits tumor growth and cancer and that melatonin, which is banned in the United Kingdom, strengthens the body's immune system. DHEA, dehydroepiandrosterone, which is not approved by the British National Formulary or the US Food and Drug Administration, is advertised as effective in preventing or curing cancer, obesity, diabetes, Alzheimer's disease, aging, osteoporosis, high cholesterol, and depression.

In the United Kingdom, products advertised with medicinal claims must have a license supported by scientific evidence and companies can be penalized for violations. However, British authorities are powerless to control the advertising and sale of products that originate in other countries.

Source: Internet sees growth of unverified health claims. *BMJ.* 1996;313:381.

Questions for Discussion

1. How can the Web be used to prevent the marketing of drugs across national borders? Should society even try to control this marketing?
2. Should Web-based marketing and sale of dietary supplements be regarded as similar to the marketing and sale of approved pharmaceuticals?

Case 3.6: Laetrile

Description

Laetrile was banned by the FDA more than 20 years ago. Yet Christian Brothers Contracting Corporation sold laetrile, also known as amygdalin and vitamin B-17, via the Web, promoting their products on multiple Web sites including CancerAnswer and ApricotsFromGod. Tablets of vitamin B-17 were available for $95 per 100 and a 21-day supply of injectable laetrile was sold for $750. The sites offered patient testimonials claiming that laetrile cured their cancer. Some patients stated they were hedging their bets by undergoing standard treatment at a cancer center and taking laetrile as well.

The FDA investigated the Web site and sent warning letters several times over a period of 2 years. An injunction was filed in the US District Court of Brooklyn, NY, to shut down the site. A federal judge issued a preliminary order barring this corporation from continuing to sell laetrile or any of its derivatives, but it is still legal to sell laetrile as a nutritional supplement in many states.

Source: Lagnado L. Laetrile makes a comeback on the Web. *The Wall Street Journal.*
    March 22, 2000: B1, B4.

Questions for Discussion

1. Should governments be given the responsibility of monitoring Web sites that make health claims and seek injunctions against sites that make unproven or fraudulent claims?

2. Would adequate investigations of the medical claims on Web sites be too lengthy and complex to be effective in curbing false or misleading claims?

3. Do consumers have the right to try alternative therapies like laetrile if they wish? If so, isn't the World Wide Web a superb medium for them?

Case 3.7: Fears of Aspartame

Description

Patients are asking many primary care physicians about the dangers of using aspartame, a sugar substitute. Allegations have been posted on the Internet linking aspartame to a number of illnesses including "Desert Storm Syndrome," fibromyalgia, seizures, cognitive decline, relapses of multiple sclerosis, blindness, migraines, and tumors, frequently lumping all these together as "aspartame syndrome." Several physicians have voiced concerns on the Internet, apparently basing their own concerns on the Internet allegations.

There is no scientific evidence that even large doses of aspartame are dangerous to adults or children. Aspartame has been heavily studied and is one of the safest food additives. When consumed, it is metabolized into formic acid and methanol in amounts that are less than those generated by eating many fruits.
Source: Squillacote D. Not-so-sweet nothings about aspartame. *Consultant.* May 1999:1323.

Questions for Discussion

1. This case is the obverse of those in which a purported remedy is inappropriately touted. Here the intent is to engender fear. Keeping in mind that no one will be harmed by not using aspartame, is using the Web as a means of spreading fear the same as using it to make false claims?

2. What would be the best way to inform the public about false claims concerning food additives, dietary supplements, and herbal products? Would it be worth the trouble? Does it matter if a false claim is positive or negative?

Case 3.8: Beware of Dangerous Medical Advice Online

Description

According to Quackwatch, an organization formed to "combat health-related frauds, myths, fads, and fallacies,"and which includes among its projects improving the quality of heath information on the Internet, "The best way to avoid being quacked is to reject quackery's promoters."

Yet phony treatments, bogus potions and ridiculous remedies are hustled with great vigor on Internet discussion groups. Pick a malady or an organ—arthritis or cancer, prostates or skin disease—and one can often find a discussion group or chat room devoted to it and used by promoters of untested and unproved compounds and devices.

A transcript from such a targeted discussion group constitutes a case study in which the rights of free speech and commerce should be weighed against the possibility of exploitation and risks to individual and public health.

Source: Barrett S. How to spot a "quacky" Web site. Quackwatch. Available at: http://www.quackwatch.com/01QuackeryRelatedTopics/quackweb.html. Accessed October 2, 2001.

Questions for Discussion

1. Does Internet information at variance with information provided to patients by their physicians and nurses have the potential to create doubt and distrust between patients and health professionals? How should this be addressed?
2. How can the consumer be protected against quack remedies that may not only be worthless but may have harmful side effects?
3. Should online support groups attempt to prevent the promotion of dangerous or useless medical advice?

Conflicts of Interest

Case 3.9 Web Site Sponsorship

Description

Drug companies are investing heavily to create a dominant presence on the World Wide Web, providing grants as high as $300,000. HealthTalk Interactive Inc's Web site, www.healthtalk.com, features panel discussions concerning new therapies. HealthTalk's hepatitis C site is funded by Amgen Inc. which produces Infergen, a brand of interferon used to treat hepatitis C. The Web site features a testimonial by a Texas patient who controls her infection with Amgen's product. Grants from Glaxo, Genentech, Amgen, and Berlex Laboratories fund patient education on other sites that deal with diseases for which these companies manufactures drugs. Similarly, Bristol-Myers Squibb's oncology drug division supports Healthology.com's cancer awareness program. The *Journal of the American Medical Association's* HIV/AIDS information site is supported by Glaxo Wellcome.

The FDA and FTC are monitoring these sites to determine if the information is tainted by commercial sponsorship and whether sponsors dictate content. There are no laws at present governing the content of health-related Web sites.

Source: Chase M. Do sponsors sway health Web sites? *The Wall Street Journal.* February 8, 2000: B7.

Questions for Discussion

1. Corporate sponsorship of health news and education has long raised questions about conflicts of interest, potential conflicts of interest, and the

appearance of conflicts of interest. How does the practice of private companies funding health information Web sites blur commercial content
and independent professional evaluation of drugs and services?

2. What role, if any, should government agencies play in regulating sponsorship of health-related Web sites?
3. If Web sites are obligated to disclose their commercial sponsorship to
   patients who access the site, how much disclosure is adequate and how
   often should it be made?

Case 3.10: Promoting Products and Services?

Description

The dr.koop.com Web site consisted of 80,000 electronic pages of information on medical conditions and advice on topics such as purchasing health
products and services. The site featured a variety of health products and services, including prescription and nonprescription drugs, vitamins, nutritional
supplements, and health insurance. Like many other commercial health Web
sites, the site accepted paid advertising.

A group of investors, including C Everett Koop, MD, the former US
Surgeon General, started the Web site and took it public in 1998 with an
initial stock offering. Dr Koop was chairman of the corporation, but executives and staff managed day-to-day operations. The company's stated goal
was to "establish the drkoop.com brand so that consumers associate the
trustworthiness and credibility of Dr C Everett Koop with our company."

In return for listing products and services, Dr Koop was originally entitled to receive 2% of the revenues derived from sales of products and up to
4% of revenues derived from sales of new products. After criticism that the
Web site blurred the distinction between objective information and advertising, the company announced that Dr Koop would no longer receive a
percentage of sales of health services or products that resulted from the
Web site.

Source: Noble HB. Hailed as a Surgeon General, Koop criticized on Web ethics. *The New
York Times.* September 5, 1999: A1, 18.

Questions for Discussion

1. Would the former Surgeon General of the United States have a conflict of
   interest if he received a commission on products and services promoted
   on his Web site?
2. Did Dr Koop have a responsibility at the outset to disclose his financial
   ties to the medical services and products featured on his Web site?
3. Since the Web site also accepted advertising for health care products and
   services, did it have a responsibility to keep the distinction clear between advertising and health education? How is a Web site like or unlike
   a newspaper or radio or television station in terms of the relationship
   between news and advertising?

Case 3.11: Listing for Sale?

Description

One feature of the drkoop.com Web site is the Community Partners Program. A list of 14 hospitals were described as "the most innovative and advanced health care institutions across the country." The Web site did not mention that the institutions on the list paid about $40,000 each to be listed.

Critics pointed out that such practices do not make clear whether information consumers receive is objective and unbiased or promotional and slanted. In response, the Web site was updated to say that the hospitals on the list "represent prominent health care institutions across the country" and to note that the hospitals had paid a fee to be included on the list.

Source: Noble HB. Hailed as a Surgeon General, Koop criticized on Web ethics. *The New York Times.* September 5, 1999: A1, 18.

Questions for Discussion

1. What went wrong? How could a prestigious and well-funded enterprise make the fundamental mistake of selling "endorsements" and failing to disclose this practice? Does the Web engender more of this kind of error than other media?
2. How do the distinctions among "conflict of interest," "potential conflict of interest," and "appearances of conflict of interest" apply in the world of Web-based health sites?

Case 3.12: Headache Resources on the Internet

Description

An increasing number of patients use the Internet to search for headache-related information. Major pharmaceutical manufacturers have created Web sites to promote their products. For example, the Glaxo Wellcome Migraine Relief Center gives information on migraine triggers and symptoms and offers a free diagnostic screening, but the only pharmacologic product mentioned on the site is Imitrex, a drug manufactured by Glaxo Wellcome. The site does not provide information about prevention of migraines or about other manufacturers' products.

Source: Genzen JR. The Internet and migraine: headache resources for patients and physicians. *Headache.* 1998;38:312–214.

Questions for Discussion

1. Is there anything wrong with a commercial Web site promoting exclusively its own product?
2. Should Web sites hew to different or higher standards than other media? That is, because the World Wide Web is more cognitively compelling

than other media, should it users be regarded as unusually vulnerable to its promotions?

Case 3.13: Professional Associations and For-Profit Web Sites

Description

The American Medical Association and 6 specialty physician groups created a company, Medem Inc, to operate Web sites and provide other physician services. The founding groups included the American Academy of Pediatrics, the American Academy of Ophthalmology, the American College of Obstetricians and Gynecologists, the American Psychiatric Association, the American Society of Plastic Surgeons, and the American Academy of Allergy, Asthma, and Immunology. The American College of Physicians chose not to participate in the venture.

Medem operates the Web site *YourPracticeOnline*. The site provides physicians with customized Web pages, secure e-mail connections with patients and other physicians, patient education materials, and news releases about medical developments. Originally physicians were to pay $70 per month per practice for the service, but they could receive it free if they agreed to have their Web site hosted by a commercial sponsor. Most physicians chose the commercial sponsorship option. Later, however, the policy was changed, and the service is now free to any physicians who are members of the participating professional organizations.

Source: Carrns A. Move over, drkoop.com: AMA launches for-profit Web venture. *The Wall Street Journal.* October 28, 1999, B4.

Questions for Discussion

1. Under what circumstances is it appropriate for services provided by medical societies to be supported by commercial sponsors? Why?
2. What do you think of Medem's decision to change the policy to make the services free to members of Medem-affiliated medical societies?

Note: This chapter is adapted with permission from Anderson JG. Health information on the Internet: Let the viewer beware (caveat viewor). MD Computing. *2000;17:19–21.*

References

1. Taylor H. Explosive growth of "cyberchondriacs" continues. The Harris Poll #47, August 5, 1999. Available online at: http://www.harrisinteractive.com/harris_poll/index.asp?PID=117. Accessed October 2, 2001.
2. McClung HJ, Murray RD, Heitlinger LA. The Internet as a source for current patient information. *Pediatr.* 1998;101:E2.
3. Impicciatore P, Pandolfini C, Casella N, Bonati M. Reliability of health informa-

tion for the public on the World Wide Web: systematic survey of advice on managing fever in children at home. *BMJ.* 1997;314:1875–1879.

4. McKinley J, Cattermole H, Oliver CW. The quality of surgical information on the Internet. *J R Coll Surg Edinburgh.* 1990;44:265–268.

5. Stolberg SG. Trade agency finds Web slippery with snake oil. *The New York Times.* June 25, 1999: A16.

6. Lagnado L. Laetrile makes a comeback on the Web. *Wall Street Journal.* March 22, 2000: B1,B4.

7. Levit K, Cowan C, Lazenby H, et al, and the Health Accounts Team. Trends: health spending in 1998: signals of change. *Health Affairs.* 2000;19:124–132.

8. Chase M. Do sponsors sway health Web sites? *The Wall Street Journal.* February 8, 2000: B7.

9. Goldsmith J. How will the Internet change our health system? *Health Affairs.* 2000;19:148–156.

10. Schwartz J. Online health sites to unveil standards. The Washington Post. May 7, 2000:A12.

11. Health Information Technology Institute, Mitretek Systems. http://mitretek.org/. Accessed August 16, 2000.

12. Health On the Internet Foundation. HON Code of Conduct (HONcode) for Medical and Health Web Sites. http://www.hon.ch/HONcode/Conduct.html/. Accessed August 16, 2000.

13. Silberg WM, Lundberg GD, Musacchio RA. Assessing, controlling, and assuring the quality of medical information on the Internet. *JAMA.* 1997;277:1244–1245.

14. Wyatt JC. Commentary: measuring quality and impact of the World Wide Web. *BMJ.* 1997;314:1879–1881.

15. Jadad AR, Gagliardi A. Rating health information on the Internet: navigating to knowledge or to Babel? *JAMA.* 1998;279:611–614.

16. Health Canada Online. Available at: http://www.hc-sc.gc.ca/. Accessed August 16, 2000.

17. National Library of Medicine. MEDLINEplus Health Information. Available at: http://medlineplus.gov/. Accessed August 16, 2000.

Further Readings

Bader SA, Braude RM. "Patient informatics" creating new partnerships in medical decision making. *Acad Med.* 1998;73:408–411.

Boberg EW, Gustafson DH, Hawkins RP, et al. CHESS: The comprehensive health enhancement support system. In: Brennan PF, Schneider E, eds. *Information Networks for Community Health.* New York: Springer; 1997:171–188.

Brennan PF, Ripich S, Moore S. The use of home-based computers to support persons living with AIDS/ARC. *J Community Health Nurs.* 1991;8:3–14.

Brennan PF, Schneider SJ, Jornquist E, eds. *Information Networks for Community Health.* New York: Springer; 1997.

Brody JE. The health hazards of point-and-click medicine. *The New York Times.* August 31, 1999:D1.

Chase M. A guide for patients who turn to the web for solace and support. *The Wall Street Journal.* September 17, 1999:B1.

Cimino JJ, Socratous SA, Clayton PD. Internet as clinical information system: Application development using the world wide web. *J Am Med Inf Assoc.* 1995;2: 273–284.

Conroy C. Care takers. *CompuServe Magazine.* Feb. 1994:10–19.

Detmer WD, Shortliffe EH. Using the Internet to improve knowledge diffusion in medicine. *Commun ACM.* 1997;40:101–108.

Ferguson T. *Health Online.* Reading, Mass: Addison-Wesley Publishing Co, 1996.

Ferguson T. Consumer health informatics. *Healthcare Forum J.* January/February, 1995:28–33.

Ferguson T, Carrell S, eds. *Consumer health informatics: bringing the patient into the loop, the proceedings of the First National Conference on Consumer Health Informatics.* Auston, TX: MailComm Plus; 1993:1–6.

Foubister V. Developing rules for the web. *Am Med News.* July 31, 2000:1–6.

Frisse ME, Kelly EA, Metcalfe ES. An Internet primer: resources and responsibilities. *Acad Med.* 1994;69:20–24.

Glowniak JV. Medical resources on the Internet. *Ann Intern Med.* 1995;123:123–131.

Glowniak JV, Bushway MK. Computer networks as a medical resource: accessing and using the Internet. *JAMA.* 1994;271:1934–1939.

Goldwein JW. Internet-based medical information: time to take charge. *Ann Intern Med.* 1995;123:152–153.

Hahn H, Stout R. *The Internet Complete Reference.* New York: McGraw-Hill; 1994.

Hayward RSA, Gagliardi A, Jadad AR. Healthcare on the Internet. *Health Measures* September 1997:28–36.

Health on the Net Foundation. Available at: http://www.hon.ch/HONcode/. Accessed: August 17, 2000.

Hi-Ethics, Health Internet Ethics. Available at: http://www.hiethics.com/. Accessed: August 17, 2000.

Hoffman M, Hock C, Muller-Spahn F. Computer-based cognitive training in Alzheimer's disease patients. *Ann NY Acad Sci.* 1996;777:249–254.

Hubbs, PR, Rindfleisch TC, Godin P, Melmon KL. Medical information on the Internet. *JAMA.* 1998;280:1363.

Jadad AR, Gagliardi A. Rating health information on the Internet: navigating to knowledge or to Babel? *JAMA.* 1998;279:611–614

Kane B, Sands DZ. Guidelines for the clinical use of electronic mail with patients. The AMIA Internet Working Group, Task Force on Guidelines for the Use of Clinic-Patient Electronic Mail. *J Am Med Inform* Assoc. 1998;5:104–111.

Kassirer JP. The next transformation in the delivery of health care. *N Engl J Med.* 1995;332:52–53.

Kehoe BP. *Zen and the art of the Internet: a beginner's guide,* 3rd ed. Englewood Cliffs, NJ: Prentice Hall; 1994.

Kocht T. Health Web sites get better at explaining complex medical data. *The Wall Street Journal.* July 14, 2000: B1.

Krol E. *The whole Internet user's guide and catalog,* 2nd ed. Sebastopol, Calif: O'Reilly; 1994.

LeBourdais E. When medicine moves to the Internet, its legal issues tag along. *Can Med Assoc J.* 1997;157:1431–1433.

Levine JR, Baroudi C. *The Internet for dummies.* San Mateo, Calif: IDG Books Worldwide; 1993.

Lindberg J.D. Providing reliable medical information to the public-caveat lector. *JAMA.* 1989;262:945–946.

Madara EJ, Ferguson T. Finding or forming your own self-help network (face-to-face or online). *The millennium whole earth catalog.* San Francisco: Harper San Francisco; 1994:172.

Medical help on the Internet. *Consumer Reports.* 1997;62:27–31.

McKinney WP, Wagner JM, Bunton MS, Kirk LM. A guide to mosaic and the World Wide Web for physicians. *MD Comput.* 1995;12:109–114,141.

McTravish FM, Gustafson DH, Owens BH, et al. CHESS (comprehensive health enhancement support system), and interactive computer system for women with breast cancer piloted with an underserved population. *J Ambul Care Manage.* 1995;18:35–41.

Mossberg W.S. Untangling the web. *Smart Money.* April 1995:127–128.

Nicoll NH. Grateful med: an easy to use information tool. *Reflections.* 1992;18:40–41.

Pannen M. Guide to the Internet: the world wide web. *BMJ.* 1995;311:1552–1556.

Parker-Pope T. Six steps to help you start your own search for medical answers. *The Wall Street Journal.* August 11, 2000: B1.

Patterson TL, Shaw WS, Masys DR. Improving health through computer self-help programs: Theory and practice. In: Brennan PF, Schneider E, eds. *Information networks for community health.* New York: Springer; 1997:219–246.

Pfaffenberger B. *The USENET book: finding, using, and surviving newsgroups on the Internet.* Reading, Mass: Addison-Wesley; 1995.

Rogers A. Good medicine on the Web. *Newsweek.* August 24, 2000;60–61.

Ruch MG. Health online. *Natural Health.* July/August 1994:78–81.

Silberg WM, Lundberg GD, Musacchio RA. Assessing, controlling and assuring the quality of medical information on the Internet. *JAMA.* 1997;277:1244–1245.

Slack WV. When the home is also the clinic. *MD Comput.* 1996;13:465–468.

Smyth KA, Feinstein SJ, Kacerek S. The Alzheimer's disease support center: Information and support for family caregivers through computer-mediated communication. In: Brennan PF, Schneider E, eds. *Information networks for community health.* New York: Springer; 1997:189–203.

Taenzer P, Zendel I, Birdsell JM, McGregor SE, Freiwald S. Cancer, me? Health care advice goes online. In: Brennan PF, Schneider E, eds. *Information networks for community health.* New York: Springer; 1997:205–217.

TRUSTe privacy statement, Cupertino, Calif. Available at: http://www.truste.org/TRUSTe_privacy.html/. Accessed: August 17, 2000.

William D. et al. Using the Internet to improve knowledge diffusion in medicine. *Commun ACM.* 1997;40:101–108.

Wyatt JC. Commentary: Measuring quality and impact of the World Wide Web. *BMJ.* 1997;314:1879–1881.

4

Privacy and Confidentiality

Illustration by Dave Harbaugh

*"He's not learning to play the harmonica . . .
It's just confidential when he discusses the
patient's medication on his cell phone."*

Reproduced from *Briefings on Health Information Security*, December 1999, p. 10. ©
2000 Opus Communications, Inc. a division on HCPro, 200 Hoods Lane, Marblehead,
MA 01945, 781/699-1872. http://www.hcpro.com. Used with permission.

More than half the hospitals in the United States are implementing electronic
patient record systems [1]. The National Research Council [2] has estimated that
the health care industry spends as much as $15 billion per year on information
technology, an amount that is expected to grow by 20% per year. By 1998, there
were 35 publicly traded health-information-technology companies with market

capitalization of more than $25 billion [3]. The importance of collecting, electronically storing, and using health information is undisputed. Consumers need it to make informed choices; clinicians need it to provide appropriate quality clinical care; and health plans and others need it to assess outcomes, to control costs, and to monitor quality [4]. However, the collection, storage, and communication of a large variety of personal patient data present a major challenge. How can we provide the data required by the new forms of health care delivery while protecting the privacy of patients? Ongoing debates concerning medical privacy legislation, software regulation, and telemedicine suggest that this challenge will not be easily resolved [5].

The problem is systemic and arises out of the routine use and flow of information throughout the health industry [2]. Health care information is primarily transferred among authorized users. The information is used not only for patient care and financial reimbursement, but also for medical, nursing, and allied health education, for research, social services, public health, regulation, and litigation, and for commercial purposes such as the development of new medical technology and marketing. The main threats to privacy and confidentiality arise from within the institutions that provide patient care or that have access to patient data for secondary purposes. By one estimate, 85% of all computer security problems involve organizational employees [6].

The purposes of this chapter are to demonstrate some of the major threats to the security of patient data and to illuminate health information policy issues arising from the dramatically increasing role of information technology in the delivery of health care. The case studies will highlight issues for which a public debate is critical to ensure that policy and legislation promote the use of information technology in ways that enhance health care but do not retard innovation.

Health Information: A New Commodity

The growth of managed care in the United States as well as changes in reimbursement rates for health services have resulted in the industrialization of the US health care system [7]. Applications of management techniques such as continuous quality improvement (CQI), benchmarking, and outcomes measurement require sophisticated information technology. These changes, combined with the development of integrated delivery systems, have provided an impetus for the development of technology for management of electronic medical records (EMRs). In both the United States [8] and Europe [9], the demand for EMRs has grown. At one level, health data are needed to manage patient care, to support research and public health activities, and to create population databases. At another level, patients increasingly use data from EMRs to seek health information, to access their own medical data, to communicate with providers, and manage chronic diseases.

The demand for sophisticated information technology in health care has spawned the growth of a large commercial health information technology industry with sales of $15 billion worth of technology in 1997. By 1995, 56% of hospitals in the United States were investing in EMRs [10]. However, much of the technology that has been developed by private industry has technical problems [11], is resisted by physicians [12], and/or raises difficult policy issues pertaining to security, privacy, and confidentiality. These policy issues must be addressed if the developing health information infrastructure is to balance the concerns of the individual against the needs of health care providers, researchers, and public health organizations.

Public Concern about Security

Protecting the privacy and confidentiality of individual health information is a critical issue. A 1993 Health Information Privacy Survey found that 27% of the respondents believed their medical information had been disclosed to others without their consent, and 24% of the health care providers surveyed reported that violations of patients' privacy had occurred within their own organization [13].

Concern for privacy has grown. A 1996 privacy survey found that 18% of the public felt that use of patient records for medical research without the patient's explicit permission was inappropriate, and, moreover, 75% of respondents felt that use of prescription data to detect fraud was unacceptable [13]. In addition, one Louis-Harris survey found that 80% of respondents believed that they have no control over their own information [14]. One third of the medical professionals surveyed said that medical information is often given to unauthorized persons.

Disclosures of significant violations of confidential medical information have spurred public concern. In Indianapolis, the medical records of patients of a psychiatrist who treated sexual problems were inexplicably posted on a web site accessible to the public. These records contained identifiable information such as names, addresses, and telephone numbers as well as intimate detail of patients' sexual problems [15]. Breaches of confidentiality have also occurred in major medical plans. Once the Harvard Community Health Plan, a major HMO, maintained medical records, accessible by all clinical employees of the plan, that contained detailed notes from psychotherapy sessions [16]. At the University of Michigan Health System, patient records could be accessed through a publicly available search engine until this security breach was discovered [17].

Patient distrust leads to the avoidance of medical care or to circumvention to avoid entering sensitive medical information into the medical record [18]. To protect their privacy, some patients pay for their own care, visit multiple providers so that there is no central repository containing their medical records, and withhold information or lie about sensitive matters (e.g., mental health services). Some patients frequently pay psychiatrists and psychologists in cash to prevent information concerning their mental health from being provided to their employers and insurance companies. The 1993 privacy survey mentioned earlier revealed that 11% of the public had not filed an insurance claim, and 7% had decided not to see a health care provider, for fear that disclosure of their health information might hurt their job prospects or their ability to obtain insurance coverage [13]. In 1995, the Harris privacy poll reported that almost 60% of the public had at some time refused to provide information to a company or business in order to protect their privacy. If patient distrust leads to a lack of reliable information, the value of data that enters an EMR will be undermined, and diagnosis and treatment decisions may be compromised.

Violations of Security by Authorized Users

Violations of the security of medical records by authorized users have always posed major problems. One study found that no fewer than 100 individuals had legitimate access to each patient's paper chart in a hospital setting [19]. These authorized accessers include physicians, nurses, medical and nursing students, pharmacists, technicians, social workers, financial managers, medical record clerks, quality assurance personnel, and billing clerks. Once patient information is electronically stored, the risk of compromising patient confidentiality increases dramatically. Not only can patient information be accessed rapidly, but it can also be linked to data from other sources. One example is the joint effort by Equifax, the largest credit report company, and AT&T to link computers in doctors' offices, hospitals, medical laboratories, pharmacies, nursing homes, and insurance companies [20]. The project envisions a medical data bank for the entire nation that would make patients' medical information available to health care providers and insurance companies. Critics point out that the resulting data bank could put the privacy of millions of patients at risk.

Privacy violations and security threats arise within the institutions that provide patient care because of mistakes, curiosity, or for other personal or financial reasons. For example, sensitive patient information may be mistakenly posted on Web sites accessible to the public. While researching an unrelated story using public search engines, a reporter from an Ohio newspaper stumbled across the sexual and medical histories, work and home telephone numbers, e-mail addresses, and credit card numbers of hundreds of patients of an Indiana psychiatrist who was also a certified sex therapist and sexologist [15]. The sensitive information, believed to have been on the Internet site for several months, included answers to questions about intimate details of the patients' sexual activities and alcohol consumption. The files had been inadvertently posted on the public Internet site. In another instance, patient data at the University of Michigan Health System was accessed through a publicly available search engine. An individual doing a search for an HMO on the university's Web site discovered the security breach [21].

When medical care is provided for celebrities, curiosity may provide the motivation for internal violations of privacy. In one such case, the medical team caring for a local celebrity at a major hospital became concerned about the level of interest in its patient's well being because the patient's hospitalization had been widely reported in the media [22]. The unit was receiving a large number of telephone calls inquiring about the patient's condition. Several of the staff members noted that a large number of individuals, some of whose names they did not recognize, had accessed the patient's electronic medical record, which contained the patient's personal information, laboratory results, and medications, as well as detailed nursing, social work, and physical and occupational therapy notes. The unit staff contacted the hospital's information services department and requested that the audit trail of their patient be examined. After staff members assigned to the unit (but not necessarily to the patient) were electronically excluded, the service's nurse manager was able to identify approximately 50 individuals not directly involved in the patient's care who had accessed the chart. Further review identified 5 to 10 additional inappropriate accessions of the patient's record during the remainder of the patient's hospitalization.

Money may provide the motivation for other internal privacy. One physician sold patient records to car dealers violations [22]. In another instance, medical students sold health records to malpractice attorneys [23]. State employees in Maryland sold patient information from the state's medical database to HMOs [24]. The London *Sunday Times* reported that anyone's electronic medical record could be obtained for £ 200 [25].

Even mischievous violations of privacy may have serious consequences. A 13-year-old daughter of a hospital employee in Florida used her mother's identification and password to access emergency room patient records [26]. She then called 7 patients and told them that they had tested positive for HIV. One teenage patient, erroneously told that she was pregnant and HIV-positive, tried to commit suicide.

Secondary Uses of Health Information

Another problem arises through the secondary use of patient-specific health information by non-health care providers. Third-party payers use these data to process claims and to manage pharmacy benefit programs. Self-insured employers are legally entitled to access to their employees' health information under provisions of the Employee Retirement Income Security Act (ERISA).

Employers argue that access to medical data is essential to manage their health benefit plans, control costs, prevent fraud, and ensure quality care. In the United States, over one third of the insured population under 65 years of age are covered by self-financed plans. Life insurers have access to beneficiaries' health information.

One major source of medical information for almost all insurance companies is the Medical Information Bureau (MIB). The MIB, a not-for-profit association headquartered in Massachusetts, was formed to exchange information among 750 member companies selling individual life, health, and disability insurance in the United States and Canada [27]. It maintains computerized medical summaries on more than 12 million policyholders.

When someone applies for life or medical insurance, member companies submit a report of any conditions that the applicant has that are related to their health or longevity. The data bank includes information about the applicant's driving record and participation in hazardous sports as well as medical data. Moreover, when an individual applies for insurance, he or she is asked to sign a form that authorizes any physician, hospital or clinic, insurance company, or other organization that has any health-related information to share that information with the MIB. The MIB does not investigate or attempt to verify any information submitted.

Once data are transferred out of the institution that provides patient care, there is little control over its use. One study found more than 200 instances in which employers and insurance companies had used information from genetic tests to discriminate against applicants [28]. In one case, a 24-year-old woman was denied life insurance when an insurance company learned that her family had a history of Huntington's disease, even though the woman herself had not been tested for the disorder. Another case involved a refusal to hire a 53-year-old man who had a history of hemochromatosis, even though the condition was asymptomatic. A second study found that one half of the Fortune 500 companies had used medical information to make employment decisions [29].

The development of databases that link patients' health information to other sources of patient-specific data poses additional risks to privacy and confidentiality. The Congressional Office of Technology Assessment [30] has pointed out that the sale of personal information from databases compiled by both government and private agencies is widespread. Data mining companies collect, analyze, and store individual-specific data, which is then sold to businesses and companies for decision making and marketing. One of the largest of these companies, Acxiom, claims to have information on 95% of US households including names, income, education, buying habits, and credit card use. A variety of other firms sell software that can be used for data mining [31].

In one case, a woman in Ohio received sexually explicit mail from a convicted rapist who was imprisoned in a Texas state prison [32]. He had obtained personal information about the woman while entering information into a data bank for Metromail Corporation. This company, a leading provider of direct marketing information to client corporations, supplies names, addresses, personal characteristics, and health-related information to clients such as direct marketers, bill collectors, and newspaper reporters.

Commercial uses of health information raise troubling issues. In 1998, a series of articles disclosed that several chain pharmacies were selling patient-specific information on prescriptions to a number of drug companies including Glaxo Wellcome, Warner-Lambert, Merck, Biogen and Hoffman-LaRoche [33]. These companies were using this information to encourage patients to refill their prescriptions with their proprietary drugs. A second case arose out of the merger of Medico Containment Services Inc, a prescription drug benefit company, Merck & Company Inc [34]. Medical information obtained by Medico in filling prescriptions for patients was used by the pharmaceutical company to try to convince physicians to switch the patient's prescription to a Merck product.

The potential profits from the development of databases have attracted a number of major corporations. As described earlier, in Equifax, the largest credit report company, and AT&T were collaborating on a project to link patient data from doctors offices, hospitals, medical laboratories, pharmacies, nursing homes, and insurance companies [35]. This project when completed would create a national database that would make patient-specific medical data available to all health care providers and insurance companies.

More recently, controversy has arisen concerning the practice of DoubleClick, an on-line advertising company [36]. The company sells advertising space to over 1500 Web sites, many of which offer health information, products, and services. Since purchasing Abacus Direct, a direct marketing company, DoubleClick can link databases that contain names, addresses, and consumer purchases by millions of consumers with data about Web sites that the individual visits. As a result of protests by privacy groups, the Federal Trade Commission and the attorneys general in New York and Michigan investigated DoubleClicks' data collection and use practices but concluded the company did not misuse personal information.

Other uses of patient-specific information, however, may be quite beneficial. Databases containing patient-specific data can be used to identify individuals who are at risk for specific health problems. Interventions can be designed and provided to these patients. An example is the identification of patients with a high probability of dying from heart disease [37] or from reactive airways disease [38]. Routine data collected and stored by the Regenstrief Medical Record System, a comprehensive EMR, were used to predict potential mortality among inner city patients who were seen by physicians in an academic primary care general internal medicine practice. The growth of public health informatics promises exciting and widespread benefits.

Personal Ethics and Self-Regulation

A host of new technologies, ranging from encryption to firewalls to audit trails, has been developed to provide security for medical information. However, current technologies designed to protect health care information are only partly effective within the institutions that provide health care. Their success depends heavily on policies that make explicit what is appropriate and inappropriate use of health information, and on personal ethics and self-regulation. Outside these settings, security technology is even less efficacious in preventing privacy violations and inappropriate use by secondary users. Privacy violations by authorized users both within and outside the institutions that provide health care will make it difficult to balance the growing need for health information and the privacy of patient data.

Most of the uses of computer-stored medical information are legitimate and will enhance the delivery of cost-effective health care. However, such usage requires the linking of patient-specific data from multiple sources and the transfer of these data outside the care setting. Moreover, these uses of medical information would be severely impaired if the data were closed to all but providers of care, or if specific approval by the patient was required before data were shared with third parties. The issue is one of control as well as privacy and confidentiality of medical information. Resolving the dilemma of increased demands for patient-specific medical information with concerns over privacy and confidentiality will be challenging.

Issues for Public Policy

In the United States, the Privacy Act of 1974 covers the use of personal data collected by federal agencies. This law prohibits disclosure of personal information for any use other than that for which it was collected without written consent of the individual. Individuals can initiate a civil suit for damages if their privacy rights are violated. The Freedom of Information Act of 1966 provides public access to the individual's records held by federal agencies [39]. However, most health information is collected, stored and processed by private organizations, which are not subject to these laws. Furthermore, state statutes governing health information generally do not cover secondary users even organizations that compile and sell health information for commercial purposes [30]. In the absence of comprehensive legislation governing privacy and security of electronic medical information, organizations incentives are lacking to invest money or manpower to improve security. Such efforts involve significant trade-off among the complexity of the security system, its costs, accessibility to data, and patients' privacy concerns.

Debates over the ethical, legal and political aspects of privacy and confidentiality are raging in Europe and North America. In the United States, for instance, more than 70 privacy and/or confidentiality bills were introduced in the 106th Congress [40]. It is, however, the Health Insurance Portability and Accountability Act (HIPAA, P.L. 104–191) that is altering the landscape. Now known widely by its acronym, HIPAA was initially a law to protect workers' health insurance when they moved between jobs. But the Congress used the measure to impose on itself a requirement to craft comprehensive, nationwide health privacy protections. Congress failed to do this, and so the task became the responsibility of the Department of Health and Human Services. That agency's final rule took effect on April 14, 2001, with entities covered by the law having 2 years to comply with it. The law included the following key provisions [41].

- All medical records and other individually identifiable health information used or disclosed by a covered entity in any form, whether electronically, on paper, or orally, are covered by the final rule.
- Providers and health plans will be required to give patients a clear written explanation of how the covered entity may use and disclose their health information.
- Patients will be able to see and get copies of their records, and request amendments. In addition, a history of non-routine disclosures must be made accessible to patients.
- Health care providers who see patients will be required to obtain patient consent before sharing their information for treatment, payment, and health care operations. In addition, separate patient authorization must be obtained for non-routine disclosures and most non-health care purposes. Patients will have the right to request restrictions on the uses and disclosures of their information.
- People will have the right to file a formal complaint with a covered provider or health plan, or with HHS, about violations of the provisions of this rule or the policies and procedures of the covered entity.
- With few exceptions, such as appropriate law enforcement needs, an individual's health information may only be used for health purposes.

The HIPAA debate pits those who regard the law as onerous against those who argue that it merely requires what those entrusted with personal health information should be doing anyway. Indeed, many of HIPAA's provisions are not new, and some recall earlier recommendations from the National Research Council [42, 43]. It is further likely that the HIPAA infrastructure will require ongoing refinement, especially regarding such issues as authorization for the secondary use of health information [44] and human subjects research and its review by institutional review boards [45].

Especially difficult debates accompany the use of unique identifiers to track patients. A proposal to use the Social Security number as a person's identifier for health information was met with protests. Since Social Security numbers are used by employers, banks, and mortgage companies and for many financial transactions, many feared it would be easy for others to access the individual's health information and use it for unauthorized purposes [46]. Patient rights groups also fear that such identifiers will permit the linking of data on individuals from multiple sources by data mining and warehousing organizations. Subsequently, Congress prohibited implementation of individual identifiers until Congress specifically approved the standard for them.

Differences between the United States and European Union (EU) with regard to regulations for protecting health information need to be addressed [47]. The EU Data Protection Directive of 1998 requires member states to prevent transmission of health information to countries like the United States that do not have laws providing a comparable level of privacy protection. This directive threatens to cut off any exchange of data between the United States and the EU states unless a compromise can be reached.

The Need for Public Policy

It is clear that information technology will play a critical role in the delivery of health care in the twenty-first century. Although spending on information technology (IT) for health care presently lags other industries, there were at one point 35 publicly traded companies, with a 1998-market capitalization over $25 billion, providing this technology [10]. Much of the research and early development of this technology has taken place in academic medical centers, but the widespread application of IT in other practice settings will depend heavily upon commercial efforts.

Recent developments in technology, in particular the Internet, World Wide Web, and use of personal digital assistants, are creating a communication infrastructure for electronic medical records. EMRs, coupled with advances in computing and telecommunication technologies, have the potential to dramatically improve clinical medicine and disease management, but security is a major barrier, especially for health applications that utilize the Internet. At a time when the market for EMR technology is growing 70% per year and is expected to reach $1.5 billion in 2000 [10], demands for public policy that protect the privacy and confidentiality of the individual's medical data are growing. At the same time access to patient data contained in EMRs will be essential to the future development of integrated delivery systems, community health networks, applications of telemedicine, and disease management programs. In particular, the use of the Internet to create a comprehensive health record for an individual who is treated at multiple sites is problematic as long as there are serious concerns about the privacy and security of the data [22].

EMR systems are critical to efforts to provide more cost-effective health care and to support the information needs of integrated delivery systems. However, these systems are vulnerable to inappropriate use both within and without the institutions that provide care. The systemic use of patient-identifiable health information by insurers, employers, drug companies, and commercial marketing firms provokes serious public concerns regarding the privacy and security of health information which must be addressed to derive the benefits of EMRs. Reports by the Institute of Medicine [48] and the Office of Technology Assessment [49] call for federal legislation and regulation to ensure the security of medical information without stifling innovative applications of information technologies. Unless a policy framework is developed, future developments and private investment in these technologies will only deepen the conflict between individual privacy concerns and the need for health information [50]. The present lack of uniform policies and standards creates problems for health care organizations that cross state boundaries and confusion about patient privacy rights and how to enforce them. The various types of pending legislation contain conflicting requirements that will have to be resolved. The roles of the private sector and the government in promoting technological developments and establishing standards to protect the privacy and security of health information will need to be negotiated. If all goes well, the standards under HIPAA will go a long way to wedding ethics, policy and professional practice in the management of personal health information.

Case Studies

Breaches of Security

Case 4.1: Sex Doctor's Patient Files on the Web

Description

A certified sex therapist operated a Web site to treat people for sexual dysfunction. For many of his patients, treatment was conducted entirely over the Internet, with no personal contact between the doctor and patient. One patient, a prominent attorney, stated that the major reason he preferred treatment over the Internet was to avoid face-to-face contact with a therapist and thus preserve his anonymity.

In March 1999, the sexual and medical histories of 15 women and 75 men who had consulted the doctor were inadvertently posted on a public Web site. This information, including names, addresses, telephone numbers, and responses to questions about intimate details of their sex lives, was found by a reporter searching on the Internet with a public-use search engine.

When contacted, the therapist said that he had relied on a computer consultant to provide security for the Web site. The consultant admitted that the personal information from patient records may have been available to the public for months but had no idea how the problem had occurred. The consultant also said additional safeguards would be put into place to prevent a reoccurrence of the incident.

Source: Laidman J, Woods M. Sex doctor's patient files show up on the Web. *Pittsburgh Post-Gazette*. March 28, 1999. Available at http://www.post-gazette.com/headlines/ 19990328doclist2.asp. Acccessed October 1, 2001.

Questions for Discussion

1. What would be required for the Internet to be secure enough to support online therapy?
2. Some patients choose online therapy in order to preserve anonymity. Is it likely that a patient's sexual dysfunction can be adequately treated without personal interaction with a therapist?
3. Did the therapist have an obligation to inform his patients that their private information including addresses, telephone numbers, and credit card numbers had been compromised? What other obligations might he have?

Case 4.2: Medical Records On-line at University Medical Centers

Description

An error on a Web-server at a university health system made thousands of patient records open to public scrutiny. When someone used the university's Web site search engine to search for a doctor, a link was opened to the confidential patient database. The patient information contained names, addresses, social security numbers, employment status, and records of treatment for colon cancer, renal failure, pneumonia and other illnesses.

When the error was detected, the records had been available to the public for two months, even though the medical center had installed scanning software to detect hackers who try to access the information system from outside of the institution. On being informed of the problem, administrative officials at the medical center were unsure whether the break in security warranted notifying the patients whose records were made available to the public.

Source: Wahlberg D. Patient records on Web 2 months. *The Ann Arbor News*. February 11, 1999: A1.

Questions for Discussion

1. How can currently available security software be implemented to prevent errors such as this one from occurring?
2. Is the medical center obligated to inform all patients whose records were affected of the security breach?
3. Will breaches of confidentiality such as this one result in patients withholding sensitive information from their health care provider? How can patients be reassured?

Case 4.3: Security Breach of a Hospital's Voice Mail System

Description

A hospital in Michigan uses a voice mail service that allows doctors to record and access notes concerning patient examinations, consultations, admission and discharge data, and even mental health records. The dictation service was provided by Dictaphone Corporation of Stratford, Conn., and is used by many other hospitals nationwide.

A TV station learned that a bug in the voice mail system allowed outside callers to access confidential patient records without identification or password. A spokesperson for the hospital said that the institution had taken immediate corrective action on learning of the problem, and that unauthorized persons could no longer access private patient information. The hospital is trying to find out how the problem occurred and to prevent it recurring.

Source: TechTV. ZDTV Cybercrime News Team Advances Major Story On Medical Records Security Breach. September 23, 1999. Available at http://www.techtv.com/aboutus/ pressreleases/story/0,23350,2340397,00.html. Accessed October 6, 2001.

Questions for Discussion

1. How can a health care institution best prevent errors that unintentionally permit public access to private clinical information?
2. Is the institution responsible for protecting the confidentiality of patient data obligated to inform all patients affected when a breach of security occurs?
3. Who should be held responsible for any harm to a patient that occurs because of the violation of confidentiality? The medical center? The company that supplied the voice mail system?

Case 4.4: Access to an Electronic Medical Record

Description

A large midwestern teaching hospital is deploying 1000 clinical workstations to allow access to EMRs. The EMR system includes a Web browser, but most health care providers at the hospital are not aware of its existence. Some of the faculty, wanting to provide students, residents, and staff with the rich array of quality medical references that are now available on the Web, have lobbied the hospital information systems group to install a shortcut to the Web browser on the clinical workstation's desktop that would allow users to access a home page indexing quality medical references with one click of a mouse.

Physicians want easy and fast access to medical information when they are providing patient care. However, the institution is concerned about unmonitored, unrestricted access to the Web from clinical workstations.

Source: Anonymous

Questions for Discussion

1. How important is it that the institution allow access to the Web only to individuals who have logged in and authenticated themselves so that their use of the system can be logged?
2. Security measures may make it more difficult for physicians to access medical information on the Web needed for patient care and medical education. How can the institution balance concerns for the security of clinical information with the needs of the physicians for easy access to this information?

Case 4.5: Electronic Mental Health Records Online

Description

A large HMO includes detailed information concerning the mental health of patients as part of their computerized medical record. The information, including diagnoses such as depression, alcoholism, and suicidal tendencies as well as intimate details of counseling sessions with therapists, is available to a large number of physicians and staff members who work for the HMO.

Administrative officials at the health plan said that the practice of putting detailed physician notes online was initiated to improve quality of care. Since all of a patient's medical records were in one file, clinical information was available to other caregivers than those the patient saw.

An official with the plan stated that the organization had instituted safeguards to protect patient privacy, and also indicated that patients were supposed to be notified when mental health information was included in their computer files. However, some subscribers indicated that they were not told about the policy and only learned that their online records contained notes concerning their mental health problems after the policy was questioned. One woman indicated that she would have limited the information that she provided to her psychiatrist if she had known that this information would be shared with other providers.

Source: Bass A. HMO puts confidential records on-line: Critics say computer file-keeping breaches privacy of mental health patients. *Boston Globe.* March 7, 1995, B1.

Questions for Discussion

1. Should sensitive data concerning such matters as mental health problems and HIV status be kept separate from other medical records to protect patient's privacy and confidentiality? Why? Why not? If so, does creating two or more separate medical records unnecessarily interfere with physician access to the patient's record?
2. Is putting intimate details of mental health problems in an EMR without first securing the approval of patients a breach of confidentiality? Does it matter who can access the EMRs?

3. Should patients be given the right to prevent sensitive data such as mental health problems from being included in their computerized medical record?
4. If patients know that information concerning their mental health problems such as alcohol and drug addiction, sexual dysfunction, and other psychological problems will be shared with other providers, are they likely to withhold information from their therapist? How can patients' concerns be addressed?

Case 4.6 A Case of Mistaken Identity

Description

The physician who saw a patient suffering from an embarrassing medical condition telephoned the patient's prescription for an expensive medication only prescribed for that specific medical condition to the local pharmacy. When the patient picked up the prescription, he was told that his insurance had expired. Further investigation revealed that while the name on the receipt was correct, the address was not. The prescription had been filled for another person with the same name who resided in the local area (whose insurance had actually expired). Apparently, when the prescription was filled, the pharmacist had only checked the name that was in the computer system and had failed to verify the identity of the patient. Alarmed by the error, the patient gave the pharmacy a false address, telephone number, and date of birth in order to maintain his anonymity.

Source: Pharmacy computer keys on names, mixing confidential records. *Risks-Forum Digest.* 1998;19:53.

Questions for Discussion

1. How could the computer system have been designed to prevent mistaken identities like this from occurring?
2. If the mistaken identity had not been corrected in the computer system, another patient's record would indicate erroneously that he suffered from an embarrassing medical condition. Could this information have adversely affected the patient's employment status or insurance coverage if uncorrected?
3. Is the pharmacy responsible for any harm that occurs to a patient as a result of the error?
4. What are the consequences of patients providing false or incomplete information when they seek health care?

Case 4.7: Confidential Medical Records on Used Computer

Description

A woman in Nevada bought a used computer through the Internet. A supermarket that had merged with a Salt Lake City pharmacy chain had originally leased the computer. After the merger, some of the leased computers that had been used in the pharmacies were returned to the leasing company and sold to a wholesaler in New York, who sold 21 of these computers over the Internet.

When the woman turned the computer on, she found that the hard drive still contained medical files on 10,000 residents of the Phoenix, Arizona, area. The files contained not only the names, street addresses, and telephone numbers of persons who had filled subscriptions through the store pharmacies, but information on drug and alcohol abuse and prescriptions for antidepressants and AIDS medications as well.

The woman sent a letter to the company requesting 2 new computers and 2 checks for $5000 in exchange for the hard drive containing the patient files. The company obtained a court restraining order when they learned that she was going through the files and contacting some of the people who medical information was stored on the computer.

Source: Markoff, J. Patient files turn up in used computer. *The New York Times*, April 4, 1997, A9.

Questions for Discussion

1. Is it appropriate for computers that have been used to store confidential patient data to be resold to other users? If so, what safeguards should be taken to prevent the accidental release of confidential data?
2. Who is responsible for the breach in security (i.e., the store pharmacies, the company from which they leased the computers, the online company that sold the computers)?

Case 4.8: E-Mail Goes Astray

Description

Kaiser Permanente, one of the nation's largest health insurers with 8.5 million subscribers, accidentally compromised the confidentiality of the medical information of 858 of its members. The problem occurred when a technician began sending out a large number of e-mail messages that had been backlogged while Kaiser's system was being upgraded. Some e-mail messages were sent to the wrong recipients. Members access the Web site and use the e-mail system to fill prescriptions, make appointments, and seek medical advice, and some of the messages contained names, home telephone numbers, medical account numbers, and medical advice. When the technician noticed the problem, he stopped sending out e-mails but did not notify Kaiser managers of the problem. The next morning, 2 Kaiser subscribers notified the company that they had received other subscriber's e-mails.

The following message appears on the Web site:

Your information is confidential. We are dedicated to keeping your personal health information confidential. We take many precautions to make sure others can't pretend to be you and get your confidential information from the Web site. As long as you don't give out your PIN, any confidential information you send or receive on this Web site can be seen only by you and Kaiser Permanente staff who have a "genuine business need."

The director of Kaiser's Web site indicated that once the error was discovered, Kaiser officials attempted to telephone each of the subscribers whose e-mails had been sent to the wrong person, and "We have fixed the problem."
Source: Brubaker B. 'Sensitive' Kaiser e-mails go astray. *The Washington Post.* August 10, 2000: E01.

Questions for Discussion

1. Who is responsible for the breach in confidentiality? The technician? Kaiser Permanente?
2. Will this breach of confidentiality discourage subscribers from accessing the Kaiser Web site to fill prescriptions and seek medical advice? How can subscribers be reassured that their information will be kept confidential in the future?

Inappropriate Use of Health Information

Case 4.9: Inappropriate Access to a Celebrity's Medical Records

Description

The medical team caring for a local celebrity became concerned about the level of interest in their patient's well being. The patient's hospitalization had been widely reported in the media and the unit was receiving a large number of telephone calls inquiring about the patient's condition. Several of the staff members noted that a large number of individuals, some of whose names they did not recognize, had accessed the patient's electronic medical record. At this particular institution, a patient's EMR contains a wide variety of information and is accessible by most users of the EMR system. The record contains not only the patient's personal information, laboratory results, and medications, but also detailed nursing, social work, and physical and occupational therapy notes. Physician's notes are, however, still maintained in a paper chart.

Many different categories of staff, including physicians, house staff, medical students, nurses, social workers, and quality assurance officers, are able to view a patient's chart. There are some controls as to who may write in or alter particular fields in the record, but most users of the EMR system may view all fields. Access is not restricted to individuals explicitly assigned to an individual patient or to individuals assigned to a particular unit. An audit trail is maintained that records who accessed the chart and when, but it is not routinely reviewed.

The unit staff contacted the hospital's Information Services Department and requested that the audit trail of their patient be examined. After electronically excluding the staff assigned to the unit (but not necessarily to the patient), a list of personnel who accessed the patient's record was sent to the service's nurse manager. The nurse manager identified approximately 50 individuals who had accessed the chart but who were not directly involved in the patient's care. Ongoing review identified 5 to 10 additional cases of inappropriate access of the patient's record during the remainder of the patient's hospitalization. The hospital's administration considered this activity to be a serious breach of policy requiring decisive and uniform action.

The managers found that most, if not all, of the accused employees believed that they had done nothing wrong. The employees expressed concern about the patient's well being and wanted to know first hand how the patient was doing. They felt that it would have been wrong to disclose this information to others, but they had not done that. Several employees claimed that they had legitimate reasons to enter the patient's chart. For example, they thought that the patient might be transferred to their service and wanted to be prepared for this possibility. The panel concluded that although the institution's expectation regarding patient confidentiality is communicated to the staff in several ways, these expectations were not well understood and should have been communicated more forcefully and clearly.

Source: Armand H. Matheny Antommaria, personal communication.

Questions for Discussion

1. Since the institution's policy regarding access to patient information was not clear, should employees not assigned to the unit who accessed the patient's medical record be disciplined? If so, how (e.g., warning, reprimand, demotion, or termination)?
2. Which, if any, of the reasons given by employees for accessing the medical record are legitimate?
3. Is the institution partly at fault for its lax security (e.g., open access to patient records, failure to review the audit trail on a daily basis)?
4. When, if ever, is accessing patient's charts to learn about unfamiliar procedures, drugs or the use of new technology appropriate?
5. In general, is reviewing a patient's chart, in whose care you are not currently participating, inappropriate even if the individual keeps the information confidential?

Case 4.10: Patients Told They Had AIDS

Description

The 13-year-old daughter of a hospital clerk in Jacksonville, Flordia, used her mother's identification and password to print out a list of the names and addresses of patients who were seen in the hospital's emergency department. She took the list home and called patients, telling them falsely that their tests indicated they were infected with the HIV virus.

Hospital officials learned that 7 patients who had been treated at the hospital had received prank calls and that one 16-year-old patient had attempted suicide after being contacted. A hospital spokesman indicated that all patients who had been treated in the emergency room over the weekend that the incident occurred had been called to determine whether or not they had received false information about their condition.

Police tracked the caller by checking the telephone number that appeared on a victim's caller ID and arrested the girl.

Source: Associated Press. Hospital clerk's child allegedly told patients that they had AIDS. *The Washington Post*. March 1, 1995: A17.

Questions for Discussion

1. Are their ways that the hospital can prevent this type of unauthorized access to patient information? How?
2. Should the hospital be held responsible for any harm to patients resulting from the incident?
3. Should the mother be disciplined or fired because of the actions of her daughter?

Case 4.11: Medical Records for Sale

Description

The *London Sunday Times* reported that detailed medical records for any individual could be purchased for £ 150. Private investigators advertised that, with the name, address and date of birth of an individual, they could provide a summary of anybody's medical records within 3 hours. The information is contained in electronic patient records created for over 56.5 million patients covered by the National Health Service in England and Wales. The files contain sexual, mental, and physical histories of patients, as well as details of medication prescribed and treatment provided for various medical conditions.

Employers who want to check the medical backgrounds of their employees frequently commission these agencies. Also, life insurance and private medical providers are obtaining data from the NHS in order to determine if potential subscribers are infected with the HIV virus or have a history of heart disease or other chronic conditions.

Source: Rogers L, Leppard D. For sale: your secret medical records for £ 150. *London Sunday Times*. November 26, 1995: 1–2.

Questions for Discussion

1. How can the privacy of patient records contained in a comprehensive national database be safeguarded? What safeguards should be put into place?
2. What should be the penalty, if any, for sale of information contained in private patient records?

Case 4.12: Medical Care Database Information for Sale

Description

Twelve of the biggest medical claim payers in Maryland, including **Blue Cross/Blue Shield**, MidAtlantic Medical Services Inc, NylCare, **Medicare and Medicaid**, submit private medical and psychiatric claims to the **Maryland Department of Mental Health and Hygiene**. This information is being entered into a new information system called the Medical Care Data Base, which is to be a state database of medical and psychological data on every resident of the state. Every physician, psychiatrist, psychologist, and chiropractor who practices in the state of Maryland will be required to submit the details of every patient encounter to the database.

Maryland prosecutors discovered that HMO staff were bribing Department of Social Services' employees to reveal confidential information on thousands of Medicaid patients. In some instances, individuals were enrolled in an HMO plan without their consent or knowledge. Sixteen state employees were arrested and pled guilty or were found guilty of bribery, improper use of private information, or Medicaid fraud.

Source: McMenamin B. It can't happen here. *Forbes.* May 20, 1996: 252–254.

Questions for Discussion

1. How secure will these data be in light of the earlier misuse of patient data? Are there ways to assure that these data will not be used for inappropriate purposes?
2. At present there are few restrictions on what the state can do with this information. Could private medical data be used by the state to identify potential criminals, to monitor children whose parents are suspected of child abuse, and so on?
3. Some critics of the system suggest that one purpose of the system is to control medical costs by monitoring services and procedures provided by physicians. In this way, unnecessary treatments could be identified. Is this an appropriate use of the database?
4. Is it practical to make certain patient records or portions of the medical record that contain sensitive information (e.g., mental health or drug addiction) completely off-limits to employees?

Case 4.13: Credit Card Fraud

Description

A computer programmer who worked for a physician group practice had access to computerized patient files that contained personal data including credit card numbers. The employee had helped to develop the system and had a high-level security clearance to access patient data stored in the information system.

The county sheriff's department notified the physician group that an em-

ployee had been implicated in credit card fraud. Someone had used patients' credit card numbers to purchase items over the Internet. Since only an experienced programmer with detailed knowledge about the system could have gained access to the credit-card numbers, the employee in question was notified that charges were likely to be filed against him. He was placed on paid leave while an investigation was undertaken and subsequently resigned.
Source: Keith Bauer, personal communication.

Questions for Discussion

1. Should credit card numbers, Social Security numbers, and other financial information be included in patients' clinical records?
2. What security measures could the group practice institute to prevent a problem like this from reoccurring?

Case 4.14: Banker Calls in Loans of Cancer Patients

Description

A banker who served on a state health commission had access to a computer registry of cancer patients in the state. He searched the registry for all cancer patients who lived in his part of the state and called in the loans that his bank had made to any of these patients.
Source: Gorman C. Who's looking at your files? *Time*. May 6, 1996: 60–62.

Questions for Discussion

1. Should members of the state health commission have access to individual patient records in a disease registry? Under what circumstances?
2. Should there be legal penalties for misuse of confidential patient data contained in public data banks? What constitues misuse?

Case 4.15: Misuse of Public Health Data

Description

A man who worked for a county health department downloaded to a laptop computer the names and addresses of people infected with the HIV virus and sent a copy of the list of 4,000 names to a newspaper. The state had permitted health workers to remove databases from the office and to take them home with them. The man took the laptop to a gay bar to check the HIV status of people that his friends were interested in dating. Friends were warned not to date certain people because their names were on the list. He also reportedly used the database to screen individuals who were interested in dating him. He further printed the list and sent it to an area newspaper.
Source: The Associated Press. AIDS list leak prompts probe, new HRS rules. The Miami Herald, Sept. 21, 1996, 5B.

Questions for Discussion

1. Under what circumstanced and restrictions should employees of the state public health department be allowed to remove private patient data from the office?
2. How should the department discipline the employee involved in this case?

Case 4.16: Integrated Health Care Systems

Description

Fundamental changes in health care delivery, including development of employer risk-retention plans, integrated delivery systems and managed care, require information systems that can facilitate extensive sharing of patient information from various sources and locations. While these new databases facilitate management and care of patients, they create substantial opportunities for invasion of privacy and misuse of patient data for social discrimination.

Many employers who provide health care benefits to their employees create self-insurance arrangements where the employer covers the costs of services used by employees and their dependents. Mergers among hospitals, clinics and health care provider groups and insurers are creating integrated delivery systems that consolidate many functions under one corporate structure. Many integrated delivery systems provide managed care so that a single entity assumes responsibility for treatment, cost containment and risk management.

These databases often contain large amounts of personal and sensitive information that is not directly related to most clinical care, for example, mental health records containing explicit information about an individual's sexual problems, anxiety, depression, history of substance abuse, genomic information, and family relations such as child and spouse abuse. This information may be sought by administrators to determine eligibility for mental health services, by a pharmaceutical group to manage the cost of medications, by insurers to establish preexisting conditions and by employers to make hiring and work assignment decisions.

Source: Gostin L. Health care information and the protection of personal privacy: Ethical and legal considerations. *Ann Intern Med* 1997;127:683–690.

Questions for Discussion

1. What social and psychological damage may result to the individual from unwanted disclosure of personal health information?
2. What are some of the justifications for creating integrated databases containing individual health and other types of data (e.g. efficiency, quality of care, research, etc.)?
3. How can integrated health care organizations balance their need for personal health information with the person's right to privacy (e.g., firewalls between the various functions such as employment, insurance, treatment etc.)?

Case 4.17: Patients' Files Used for Obscene Calls

Description

An orthopedic technician who had been convicted of child rape and indecent assault used the password of a former hospital administrator to gain access to confidential medical records of 954 patients at a major hospital. He then made obscene telephone calls to female patients as young as 8 and 9 years old.

The technician's access to the confidential patient records began in December and continued until he was fired 4 months later. The hospital was not aware of the problem until a trace on the telephone line of a girl who was receiving obscene calls indicated that the calls originated from the hospital. The computer system failed to detect the outdated password and did not alert employees responsible for maintaining the information system that one individual was accessing a large number of patient files. Moreover, the hospital did not conduct background checks when hiring new employees.

Source: Brelis M. Patients' files allegedly used for obscene calls. *The Boston Globe.* April 11, 1995: 1.

Questions for Discussion

1. Should health care institutions conduct background checks on new employees who will be allowed access to confidential patient information? What information should be accessible to such employers?
2. How could the hospital have prevented the misuse of patient information from occurring? Was the hospital's security system at fault for this breach of security?
3. Should the hospital be held accountable for the actions of the technician?

Case 4.18: Abortion Clinic Sues CompuServe

Description

An abortion clinic in Florida filed a federal lawsuit—later dismissed—against CompuServe (later acquired by America Online) and TML Information Services, another Internet service provider (ISP), which claims to be the largest provider of online Internet access to motor vehicle records in North America. The two ISPs were accused of permitting antiabortion activists to access databases that allow them to match the drivers license numbers of planned parenthood clinic visitors with their home addresses, telephone numbers, and other personal information. These antiabortion groups were accused of using this information to harass women and to discourage them from visiting family planning agencies.

Critics argue that these contacts by activists are forcing women to drive long distances to other communities for family planning services in order to maintain their anonymity. Moreover, they point out that the Federal Drivers Privacy Protection Act of 1994 prohibits state motor vehicle departments from disclosing personal information about an individual, unless the information is requested by law enforcement agencies, court officials, private investigators, businesses that need to verify certain information, or researchers. CompuServe, in defense, argues that they only offer access to the databases, that they did not actually collect the data, and that the Federal Telecommunication Act states that ISPs can not be held legally liable for content published on the services by third parties.

Source: Macavinta C. Abortion clinic sues CompuServe, ISP, *CNET News.com.* January 6, 1999. Available at http://news.cnet.com/news/0-1005-200-337039.html. Accessed October 6, 2001.

Questions for Discussion

1. Will patients who visit family planning agencies avoid local agencies if they fear they can be identified through online Internet services? How can they be reassured?
2. Should online service providers be held responsible for harm caused by content posted on the service by third parties? What should be the penalties, if any?
3. Should provisions of the Federal Drivers Privacy Protection Act of 1994 that prohibit state motor vehicle departments from disclosing personal information about individuals also apply to ISPs?

Case 4.19: Men Contract Syphilis after Meeting in an Internet Chatroom

Description

Up to 99 men may have contracted syphilis from sex partners they met in Internet chat rooms. One man had 47 sex partners. America Online declined to release their names for medical testing citing privacy reasons. (Different aspects of this case are addressed in Case 7.9.)

Source: Known only by computer. *Comput Med* 1999;28:1.

Questions for Discussion

1. Is America Online ethically obligated to release to public health agencies the names of the men exposed to syphilis so that they can be contacted and offered testing and treatment, if they are found to be infected?
2. Under what other circumstances, if any, should ISPs be required to provide identifying information about subscribers?

Secondary Use of Medical Information

Case 4.20: Privacy of Pharmacy Data

Description

Ms K used her pharmacy drug card to buy an antidepressant prescribed by her physician to help her sleep. By using the card, she saved $9 off the price of a month's supply of trazodone. A short time later, PCS Health

Systems, the pharmacy benefit company that manages prescription drugs for her health insurance plan, sent her physician a letter. The letter stated that Ms K employer had noted that she was using antidepressants and had enrolled her in a health prevention program called Journeys: Paths through Depression. In the future, her prescriptions would be monitored, and her doctor would be notified if she failed to renew her prescription in a timely fashion. Ms K would also be sent educational materials on the management of depression.

Since Ms K was taking the medication for sleeping problems related to menopause, she felt that she had been mistakenly enrolled in the depression program without her consent. Moreover, she feared that her company, Motorola, had labeled her as mentally ill, and that this would affect the possibility of future promotions.

Source: O'Harrow R. Plans' access to pharmacy data raises privacy issue; benefit firms delve into patient records. *The Washington Post.* September 27, 1998: A01.

Questions for Discussion

1. Patients are likely to lose trust in the medical profession when they are enrolled in health plans without their consent. How can this problem be prevented?
2. Are actions such as these by pharmacy benefit companies necessary to control the rising costs of drugs for health plans?

Case 4.21: Tracking Consumers on the Web

Description

Pharmatrak Inc, a Boston firm, tracks Internet users on behalf of pharmaceutical companies. When a consumer visits one of its clients' Web sites, Pharmatrack places a cookie on the consumer's computer that permits the company to record consumers' activity when they access Web sites maintained by 11 pharmaceutical companies, including Pfizer Inc, Pharmacia, SmithKline Beecham PLC, Glaxo Wellcome PLC, Aventis Pharmaceuticals Inc, Novartis Pharmaceuticals Corp, American Home Products Corp, Hoffmann-La Roche Inc. For example, the company can tell when a consumer downloads information about HIV or a prescription drug. Pharmatrack is a subsidiary of Sonnenreich's holding company, Global Communications Ltd, which also owns Agritrak. Agritrak provides the same services to agri-biotech companies. Pharmatrak provides monthly reports to the 11 drug companies.

Pharmatrack does not post privacy policies on client Web sites that indicate that the company collects and uses data on consumers who access them. Only the following statement on Pharmatrack's Web site indicates that the company plans to identify consumers who access their clients' Web sites:

In the future, we may develop products and services which collect data that, when used in conjunction with the tracking database, could enable a direct identification of certain individual visitors.

Pharmatrack also offers a service called Netwatcher which uses a sophisticated search engine to search the Web for any mention of client companies, their products, or company executives. Currently it searches Web sites, but in the future it will also search online chat rooms. Any negative reports on the Internet are identified so that the client company can respond and, if necessary, take legal action.

Source: O'Harrow R. Firm tracking consumers on Web for drug companies. *The Washington Post.* August 15, 2000: E01.

Questions for Discussion

1. Is Pharmatrack acting inappropriately by monitoring consumers who access Web sites of pharmaceutical companies without notifying the consumer that it is doing so? What would constitute appropriate notification (e.g., a statement posted on the Web site)?
2. Will secret monitoring of consumers' use of the Web dissuade them from using the Internet to seek health information? How can this be prevented?
3. Are the services offered by Pharmatrack comparable to the use of hidden cameras and private investigators to monitor consumer behavior secretly? If so, how do these practices differ?

Case 4.22: Release of Medical Information

Description

A self-insured employer began the process of converting employees' medical records to a computerized database. The stated purpose was to improve the efficiency of claims management. Each of 9000 employees was asked to sign the following release form:

> To all physicians, surgeons and other medical practitioners, all hospitals, clinics and other health care delivery facilities, all insurance carriers, insurance data service organizations and health maintenance organizations, all pension and welfare fund administrators, my current employer all of my former employers and all other persons, agencies or entities who may have records or evidence relating to my physical or mental condition:
>
> I hereby authorize release and delivery of any and all information, records and documents (confidential or otherwise) with respect to my health and health history that you, or any of you, now have or hereafter obtain to the administrator of any employee benefit plan sponsored by Strawbridge & Clothier, any provider of health care benefits offered or financed through a benefit plan sponsored by Strawbridge & Clothier, and any insurance company providing coverage through any benefit plan sponsored by Strawbridge & Clothier.

While some employee expressed concern about the sweeping nature of the release form they were asked to sign, only about 10 to 12 employees challenged the validity of the company's claim to such broad authorization for access to their medical records. For these employees, the company added a clause to their forms that specified that their medical records could only be used by insurance companies to process medical claims.

Source: Dahir M. Your health, your privacy, your boss. *Philadelphia City Paper.* May 28–June 4, 1993: 10–11.

Questions for Discussion

1. How much access do employers need to employees' medical records in order to administer health benefit plans?
2. Should employees be informed of their right to limit access to their medical records to insurance companies to process medical claims? Could employees be penalized by their employer if they limit access to their medical information?

Case 4.23: The Medical Insurance Bureau

Description

Mr D tried to increase his life insurance from $30,000 to $100,000 and was turned down by 3 insurance companies. He later found out that when he completed an application for life insurance, the insurance company requested a file of his medical information from the MIB. The MIB maintains computerized medical records on over 12 million American and Canadian policyholders. During a physical examination, Mr D had told his physician that he drank a couple of six packs of beer each month. Information that was submitted to the MIB erroneously stated that Mr D drank 2 six-packs of beer per day. After Mr. D's laboratory tests indicated elevated liver enzyme levels, he asked his physician if the elevated levels could have been caused by 1 or 2 drinks that he had consumed the evening before his examination or by use of marijuana. His MIB file subsequently indicated that Mr D's alcohol and drug use were significant threats to his health and longevity.

Source: Anthony J. Who's reading your medical record? *American Health.* November, 1993:55–58.

Questions for Discussion

1. Is it a violation of doctor-patient confidentiality when questions and answers that a patient raises regarding alcohol and drug use are reported in a databank and can be used against the patient by employers and insurance companies? Is it appropriate for the physician to protect the patient's privacy by not recording this information in the patient's record?
2. If private conversations between a doctor and patient are reported to a databank, will this practice erode the patient's confidence in his or her health care provider?
3. If patients fear that information they provide to their health care provider can be used against them, are they likely to withhold important health-related information from their provider? How can this be prevented?

Case 4.24: The Psychiatric Case Register

Description

In Israel, the Ministry of Health has maintained a psychiatric case register (PCR) since 1950. All inpatient psychiatric episodes for any person in the country are recorded cumulatively in the register. Proponents of the system argue that the PCR is a research tool that permits long-term data collection on large patient groups. The data contained in these databases can be used for epidemiologic studies and for preventing, treating, and controlling mental diseases. PCRs also provide important data that can be used for planning health services.

While the law requires that information regarding mental patients be kept confidential, it provides a list of public agencies authorized to obtain individual psychiatric information. The list includes the Firearms Registrar, the army, and the Institute for Road Safety. The Firearms Registrar uses information from the PCR to screen applicants for firearms permits. The army uses information from the register in screening draftees. The Ministry of Transportation uses this information in licensing drivers for special vehicles such as buses. The information is available to anyone from one of the listed agencies who is conducting an investigation, including the police, the secret services, income tax investigators, and so on.

Source: Rahav M. Labeling the mentally ill through psychiatric records: The Israel case. *Isr J Psychiatr Relat Sci.* 1985; 22:221–231; Rahav, M. The computerization of Israel's Psychiatric Case Register: Blessings and dangers of automated information systems. *Bull Royal College Psychiatr.* 1986;10:215–218.

Questions for Discussion

1. Does the special attention paid to patients with psychiatric problems reflect public fear and mistrust of the mentally ill? Will the use of the case register in this way reinforce public fears of the mentally ill?
2. Does the implementation of the PCR result in a safer and more secure society, or are public fears of the mentally ill largely groundless?
3. Is the widespread use of data from the PCR an unwarranted invasion of patients' privacy? Why, or why not?
4. Do computerized information systems like the PCR make access to patients' private medical information too readily accessible to government agencies? How can misuse of these data be prevented?

Case 4.25: Wellness Plan Files Used to Fight Job-Related Injury Claims

Description

Mr F, 54, worked for Adolph Coors Co brewery in Golden, Colorado. The company sponsors a wellness center for employees that includes a gym, medical clinic, and counseling center. Employees can work out in the gym, visit a medical clinic, and attend health promotion classes on relaxation and anger management among others. Seventy percent of the company's employees agree to complete a Health Hazard Appraisal to qualify for a 5% savings on their medical coinsurance payment. This questionnaire asks about problems with spouses and children, drinking and smoking habits, sexual difficulties, and mental health.

Detailed medical records are kept on all employees. Moreover, employees who are reimbursed for work-related ailments must sign a release that makes all of their medical records, including accounts of treatment by psychotherapists, available to the company's management. The company estimates that the wellness program saves $2 million a year by reducing sick leave and medical costs.

In 1992, Mr F died of a heart attack 2 weeks after being demoted from an office job to manual labor. His widow filed for survivor's benefits available if his death was due to job-related causes. The company used information from Mr F's medical records concerning his long-term smoking habit to persuade an administrative judge to deny the benefits.

Source: Schultz EE. Medical data gathered by firms can prove less than confidential. *The Wall Street Journal*. May 18, 1994: A1, A6.

Questions for Discussion

1. The extensive collection and use of medical information for a variety of management purposes increase the problems of maintaining the confidentiality of patient information. How can misuse of these data be prevented?
2. Is it appropriate for the company's managers to have access to all of an employee's medical information including records of sessions with psychotherapists? What information, if any, should managers have access to?
3. Should the company be permitted to use information in an employee's medical records to discredit compensation claims for work-related problems? If not, how can they be prevented from doing so?

Case 4.26: Dispute over Use of Medicaid Records

Description

In New York City, homeless and drug-addicted people are required to work in order to qualify for welfare benefits. One program is designed to identify nondisabled HIV-positive people on welfare who are eligible to participate in the Workfare program. The city plans to crosscheck Medicaid records that indicate the use of HIV medications against the city records of Workfare participants. A preliminary search of the Medicaid database revealed the names of 10,000 HIV-positive people who might be eligible to participate in the Workfare program.

Critics of the plan argue that using Medicaid records for nonmedical purposes is improper and may be illegal, since federal and state laws prohibit using confidential Medicaid records for nonmedical purposes. Others voice concerns about forcing HIV-infected persons to work in situations were they might be unable to take their medications as prescribed or may be too fatigued to carry out their assignments. Further concern is that use of the Medicaid database by the Human Resources Administration as proposed may intensify opposition to the reporting of names of newly diagnosed HIV-positive patients to the state public health department.

Source: Lerner S. Medicaid records dispute plagues city's Workfare program; A question of privacy. *Village Voice*. February 23–29, 2000. Available at: http://www.villagevoice.com/issues/0008/lerner.php. Accessed October 6, 2001.

Questions for Discussion

1. Is the use of the Medicaid database to identify persons who might be eligible for the Workfare program an appropriate use of these data? If not, what uses of the database are appropriate?
2. Can such use of the Medicaid data result in harm to persons who are HIV-positive.
3. Should the city use the Medicaid database for other purposes such as identifying people who are using illegal drugs?
4. If the city uses the Medicaid database for Workfare eligibility searches, will it discourage some people from disclosing their HIV status and seeking treatment? What other undesirable consequencies are possible?

Case 4.27: Billing Data Released in Fraud Case

Description

In Kansas City, Kansas, 2 doctors, 3 hospital administrators, and 2 lawyers have been charged with conspiracy to defraud Medicare by paying for patient referrals. Five hospitals are alleged to have been involved in the kickback scheme, and 3 others are under investigation.

The legal brief submitted by the Assistant US Attorney for Kansas includes laboratory billing records containing the names of 274 patients, the tests that were performed on each patient, the date each test was performed, and billing charges. This information is now part of the public record.

Critics said that they were astounded that private information about patients along with their names was made public as part of the legal motion by the Assistant US Attorney who is prosecuting the case. They observed that at least the names of patients should have been deleted from the brief. Some of the persons whose laboratory test data was released could lose their jobs or health insurance benefits. According to the Joint Commission on Accreditation of Healthcare Organizations, if a hospital or an accredited medical laboratory were to release information that contained patients' names or other identifiers, it would constitute a serious breach of confidentiality.

Source: Moore JD. Confidentiality casualty: Patient billing printouts released in Kansas fraud case. *Modern Healthcare*. September 14, 1998, 3.

Questions for Discussion

1. Is it necessary and/or appropriate to make the names and medical information about patients' part of the public record in legal cases involving health care providers? How can patients' privacy be protected in legal proceedings? For example, should the hospitals and medical laboratories that performed the tests have deleted patient names and other identifiers from the billing data before releasing it to the investigators?
2. Can the hospital or medical laboratory that provided investigators with the billing data be held responsible for harm to patients whose medical information was released?
3. Should there be safeguards against the release of medical information with patient names or identifiers in fraud and abuse investigations? What safeguards would be appropriate?

Data Sharing

Case 4.28: Creating a National Database

Description

Equifax, the nation's largest credit card company, and AT&T announced a joint project that will connect computers in doctors' offices, clinics, hospitals, medical laboratories, pharmacies, nursing homes, and insurance companies. The project, when completed, would create a national database that would make individual medical data available to all health care providers and insurance companies. The resulting EMR system would permit specialists to review the notes of the primary care physician; physicians could send prescriptions directly to the pharmacy; and emergency personnel could access a patient record from anywhere in the United States.

Critics point out that medical databases that permit easy access to patient data raise a new range of privacy problems. Literally thousands of persons will have access to the database. Even if security measures were enacted, there would be no way to closely monitor use of the information by a variety of users in different institutions. Furthermore, a patient has little recourse under current regulations if his/her medical data is inappropriately used.

Source: Davis R. Online medical records raise privacy fears. *USA TODAY*. March 22, 1995: 1A.

Questions for Discussion

1. Are there ways to safeguard the privacy of medical information when a common database is shared by multiple providers and institutions? How?
2. Should the consent of the individual patient be necessary before his/her medical data is included in a shared database?
3. Who should be held accountable if patient data contained in a shared

database is used inappropriately? The institution that created the database? The users of the information?

Case 4.29: Medical Web Sites Faulted on Privacy

Description

A survey of privacy practices of 21 major health Web sites was conducted by the Health Privacy Project at Georgetown University. These sites offer clinical and diagnostic information, products and services, opportunities to interact with health professionals, and the ability to create an online personal health record. The study found that, though these sites had published policies to protect privacy of individuals who access the information contained on the Web site, in general, the sites' practices did not match their stated policies. As a result, the public has an unwarranted trust of these Web sites.

Health Web sites have access to a great deal of personal information about consumers. The study found that many of these sites use "cookies", that is, code placed on the users' computer that allows the site to identify the user on a return visit. Moreover, many health and medical Web sites allow advertisers and other third parties access to users' information. Some sites ask users to fill out registration forms containing personal information in order to gain access to information. Many of the sites have arrangements with outside online advertising firms such as DoubleClick, Inc. When a user clicks on the banner of the outside advertising firm, the health Web site can track the user and pass information to a database compiled by the advertising firm. In some instances, consumer information is passed on to yet another Web site unbeknownst to the consumer. For example, when consumers completed a survey on a Web site sponsored by OnHealth Network in Seattle, they were actually logged on to a Portland, Oregon, WellMed Web site that was collecting user information. Consumers are unaware that this information can be used to personally identify them, and that it may be sold to third parties for other uses such as direct marketing.

The study found that consumers are willing to share some personal information with Web sites in return for some of their services. At the same time, 75% of people surveyed expressed concern that Web sites are sharing personal information with others without first securing the consumer's permission. Almost 20% of respondents indicated that they do not seek medical information online because of concerns about their privacy.

Source: Schwartz J. Medical Web sites faulted on privacy. *The Washington Post.* February 1, 2000: E1.

Questions for Discussion

1. Should medical Web sites be required to alert consumers that they are collecting personal data and to publicly state how these data are used? Is a notice posted on the Web site sufficient?

2. Should consumers be given the option to withhold personal information from the medical Web site they access and still be able to obtain the information they are seeking?
3. Is there any way for a consumer to determine whether a medical Web site adheres to its stated privacy policy?
4. Will violations of privacy by medical Web sites inhibit large numbers of the public from accessing the information and services provided by these sites? How can this be prevented?

Case 4.30: Privacy on Cancer Support Web Sites

Description

A number of nonprofit Web sites provide information and support for cancer patients. These sites include Oncolink, sponsored by the University of Pennsylvania, and the National Institute of Health's Medline. Several commercial Web sites also provide support and information to cancer patients.

Some sites ask patients to provide personal data in exchange for medical information and expert opinions. Personal data from cancer patients can then be sold to advertisers and business partners who, in turn, sell products to the patients. The products include online pharmaceutical services, syringes, colostomy bags, home health care services, private-duty nursing services, and floral delivery among others.

Web sites generally promise to treat health-related information as confidential but also state that any information that consumers provide or post to the site can be used for other purposes including marketing research. Some of the cancer sites clearly state that demographic data will be shared with pharmaceutical companies. One Web site warns patients that failure to provide all the requested information will limit access to information available on the site. Another offers patients a free cancer profiler service that provides personalized information on medical studies related to the patient's type of cancer. The company that sponsors the Web site allows drug companies to provide gift certificates for a premium version of the profiler for support groups and doctors.

Source: Landro L. Cancer support sites are raising questions about medical privacy. *The Wall Street Journal.* April 28, 2000: B1.

Questions for Discussion

1. When commerical Web sites offer services to patients with life-threatening illnesses such as cancer, is there a potential for exploitation of vulnerable patients?
2. Do commerical Web sites have a responsibility to alert consumers to how they intend to use personal information they collect? Is it appropriate for health-related Web sites to share patient data with third parties as long as they inform consumers that they plan to do so?
3. Do these commerical Web sites have an obligation to inform consumers of their financial sponsors?

4. Is it permissible for cancer information Web sites to refuse to provide services if the patient refuses to provide personal information?

Case 4.31: Violation of the Privacy of Pharmacy Customers

Description

Prescription databases are being used to remind customers who don't refill prescription and to market new products. CVS Corporation and Giant Foods sent confidential prescription information, names, addresses, and other personal information to Elensys Inc in Woburn, Massachusetts. Elensys arranged for drug manufacturers to pay pharmacies for the right to send educational material to customers who had specific conditions. The purpose of these mailings was to stimulate drug sales by encouraging patients to refill their prescriptions. Letters to patients were written on the pharmacy's letterhead and signed by the pharmacist. One customer, who was using a prescription nicotine replacement product to stop smoking, received a personalized letter on the pharmacy's letterhead promoting a new drug.

A class-action suit was filed against CVS, Elensys, Glaxo Wellcome, Warner-Lambert, Merck and Hoffman-La Roche. The suit allerges that the companies violated the privacy of pharmacy customers.

Source: Harrow RO. Prescription sales, privacy fears; CVS, Giant share customer records with drug marketing firm. *The Washington Post.* February 15, 1998: A01.

Questions for Discussion

1. Under what conditions, if any, should pharmacies be permitted to share prescription data with outside organizations?
2. Should pharmacies have to secure authorization from each patient before sharing prescription information? Or is it acceptable for them to post a notice of their intent to do so?
3. Is the use of such strategies appropriate when they are used for noncommercial purposes such as health risk management? Are there other uses of such strategies that are acceptable?

Case 4.32: Marketing Prescription Drugs

Description

Merck & Company, one of the world's largest pharmaceutical companies, used information from Medco Containment Services, a prescription drug benefit company, to market its prescription drugs. When prescriptions written by the patient's doctor reached Medco, a pharmacist employed by Medco would call the patient's physician and encourage the physician to switch the prescription to a drug manufactured by Merck & Co Inc. Doctors were not told that Merck owns Medco and that Medco employed the pharmacist.

The Minnesota State Attorney General and attorneys general from 16 other states challenged this practice as a violation of consumers' privacy rights. In a settlement, the two companies agreed (i) to disclose to consumers' doctors that the pharmacist is calling on behalf of Medco, (ii) to substantiate and honor any claims of cost-savings resulting from changing patients' prescriptions, (iii) to advise customers that they may contact their doctor if they do not want their prescription changed, and (iv) to advise customers about the extent to which patients' confidential information contained in the Medco database can be made available to third-parties including their employers.

Source: PRNewswire. Minnesota takes the lead on agreement to protect 41 million Americans. October 25, 1995. Available at http://www.epic.org/privacy/medical/merck.txt. Accessed April 11, 2001.

Questions for Discussion

1. Prescription drug benefit companies responsible for cost-containment frequently promote the drugs and products of their affiliated company. Should they be required to inform patients of their affiliation?
2. Do you feel that the privacy measures that Medco and Merck agreed to take are sufficient to protect the privacy and confidentiality of patients' medical data?
3. Should customers have the option to request that their prescription information not be shared with others and used for marketing purposes?

Case 4.33: Inappropriate Use of Patient Pharmacy Data

Description

A woman filled her daughter's prescription for Zoloft, an antidepressant, at a supermarket using her daughter's store discount card to pay. The card entitles the holder to discounts and also keeps track of purchases. Later, when the woman made a subsequent purchase using the card, the cashier gave her a coupon for "depression sufferers," which listed a telephone number for information about clinical depression.

The woman subsequently called the store, spoke with the senior pharmacist, and asked if the store's computerized cash register had access to customers' prescription records. It appears that individual prescription profiles were being used to target patients for marketing purposes. The woman then wrote to the physician who had written the prescription objecting to the violation of confidentiality.

Source: Pomerantz JM. A case of inappropriate use of patient pharmacy data. *Drug Benefit Trends*. 2000; 12:2,6.

Questions for Discussion

1. Is it appropriate for the store pharmacy to use confidential patient data for marketing purposes without first securing the customer's consent? Should the patient's physician be informed of this practice?
2. The linking of the store's cash register to prescription data exposes con-

fidential patient data to store employees and possibly other customers. How can this be prevented?

Case 4.34: Doctor Shaken to Find Personal Data on the Web

Description

In 1997, the California Medical Board, which licenses physicians to practice, put its database of license information on a Web site. The listings included names and home addresses. When the Union of Physicians & Dentists objected, on the basis that the inclusion of home addresses might jeopardize the safety of physicians, the board permitted physicians to change their home address to an office address or a post office box.

At a later date, a family physician in Fremont, California, provided the California Medical Board with her home address in order to receive her medical license to practice in the state. Later, while searching the Internet, the physician found her name and home address in a directory on the WebMD site. The California Medical Board had sold its data base of more than 100,000 physicians to the Pacific Business Group on Health, which in turn had sold the data to a firm called Health Pages in New York. In 1999 the California broad processed 168 requests for lists of doctors from health agencies, researchers, and marketing firms. The service costs 3 cents per name.

Source: Heim K. Doctors shaken to find personal data on the Web. *Mercury News*, April 16, 2000. Available at http://www.mercurycenter.com/svtech/news/indepth/docs/ priv041700.htm. Accessed October 6, 2001.

Questions for Discussion

1. Is it appropriate for the medical licensing board to sell information contained in its database to commercial interests? Should it first have sought permission from each individual?
2. Should physicians applying for licenses to practice in California be given an option to exclude their name and personal data from lists that are sold to third parties?
3. Is it inappropriate to collect data ostensibly for one purpose and then to use it for other purposes without informing the persons who contribute the information in advance?

Case 4.35: Data Mining

Description

Ms D, a Texas woman, received a letter containing personal details about her and threatening her with rape. The unknown writer of 12-page letter knew her birthday, her favorite magazines, the fact that she was divorced, and even her favorite type of soap.

It turned out that a convicted rapist serving time in a Texas prison had written the letter. He had obtained the information from a product questionnaire that the woman had filled out and sent back to the company in order to obtain product discount coupons and free samples. The questionnaires were sent by Metromail Corporation, a national seller of direct marketing information, to the prison, where unpaid prisoners entered the information into computer files. Metromail claims to have information on 90% of US households in their database.

Companies, like Metromail, which profit from collecting, combining, and selling personal information to be used for direct marketing and other purposes, argue that the public has no legal right to know or need to know that these data are being collected and made available for commercial purposes. When Ms D joined a class action suit against Metromail, the company used its database to generate over 900 items of information about her. This information totalled 25 spreadsheet pages and contained information on income, marital status, hobbies, brand of antacid tablets, sleeping aids, and hemorrhoid remedies she used. This information was attached to a memorandum written by an information technology systems analyst and circulated to top executives and the company's lawyers.

Source: Bernstein N. Lives on file: The erosion of privacy—A special report; personal files via computer offer money and pose threat. *The New York Times.* June 12, 1997: A1.

Questions for Discussion

1. Since Ms D voluntarily contributed personal information in a consumer survey, is she partly to blame?
2. Should companies that collect and sell personal information for marketing purposes screen the employees they hire to enter the data? Should they be permitted to use prisoners to enter data?

Rating Physicians

Case 4.36: Rating Physicians on Outcomes

Description

Newsday filed a suit under the Freedom of Information Act to obtain information about physicians in New York state who performed coronary artery bypass graft operations. The information indicated that mortality rates varied among physicians performing surgery at the same hospital, and that physicians who performed fewer than 50 bypass operations annually had higher mortality rates than physicians who performed more than 50 heart operations. *Newsday* released the names and adjusted mortality rate for each of 140 physicians to the public.

Consumer advocates argued that the public has a right to know the records of physicians who perform major surgery such as heart bypass operations so as to more reliably choose a surgeon with a better safety record. Critics argued that the providers' records should be kept confidential, and that the health care professions have policies and procedures in place to protect patients. They also argued that the release of physicians' records to the public would inhibit quality-assessment programs designed to improve the performance of their entire provider community.

Source: Freidman E. As long as it comes out all right. *Healthcare Forum Journal.* July/
    August 1992:11–15.

Questions for Discussion

1. Does the public have a right to know the adjusted mortality rate for each physician's patients?
2. Will releasing this information to the public unjustly damage physicians' reputations, expose them to malpractice suits and denial of hospital privileges, and cause them financial harm?
3. Do bad outcomes necessarily imply poor performance? For example, do outcome measures adequately adjust for age, gender, diagnosis, and severity of illness, and a host of other factors that may account for variation in outcomes?
4. Will health care providers agree to collect outcome data to use to improve performance if they fear this information will be made public?

Case 4.37: Physician Disciplinary Reports on the Internet

Description

New York and other states give consumers Internet access to doctors' malpractice and disciplinary records sometimes including information on all liability payments for each doctor, and a synopsis of allegations in each case. Doctors in some cases are to post an explanation for each case that is listed. Information on each doctor's education, board certifications, community service, health plan participation, criminal convictions, and disciplinary actions are made available on the Web site.

Source: Prager LO. Internet disciplinary reports likely for NY physicians. *Am Med News.*
    April 10, 2000. Available at http://www.ama-assn.org/sci-pubs/amnews/pick_00/
    prsc0410.htm. Accessed October 9, 2001.

Questions for Discussion

1. Will access to this information permit consumers to make more informed choices of their physicians? Or, is information about malpractice claims settlements an unreliable basis for rating physicians' performance?
2. Does the proposed bill to put physicians' malpractice and disciplinary history online indicate general dissatisfaction with the oversight provided by the state medical board? Is this dissatisfaction valid?
3. Are their more credible and reliable approaches to assist consumers in making informed choices about physicians? What are they?
4. Will doctors be reluctant to settle malpractice cases out of court if these data are made available to the public on the Internet? If so, will physicians' reluctance to settle malpractice cases result in a backlog of court cases?

Case 4.38: The Wall of Silence

Description

Ms G had a girl by caesarian section at Beth Israel Hospital in Boston. After the surgery, Dr Z her obstetrician, carved his initials on her abdomen. After Dr Z's hospital privileges at Beth Israel Hospital were suspended, he continued to practice medicine for an additional 5 months, but he subsequently lost his license to practice in the State of Massachusetts.

Ms G stated at the US House Commerce Subcommittee hearings that if she had been allowed to see Dr Z's file in the National Practitioner Databank, she would have known about other patient complaints concerning Dr Z and would have chosen another obstetrician.

The National Practitioner Databank was created by the US Congress in 1986 to inform state medical licensing boards in one state about malpractice, civil actions, and licensing board actions against physicians in other states. The databank is accessed by hospitals, HMOs, insurance companies, and state medical licensing boards, but the public has no access to information about physicians contained in the databank.

Source: CNN. Lawmakers consider opening database on doctors to public. March 1, 2000. Available at: http://www.cnn.com/2000/HEALTH/03/01/doctor.disclosures/. Accessed October 7, 2001.

Questions for Discussion

1. Should the public have access to information on individual physicians and dentists contained in the National Practitioners Databank concerning malpractice payments, disciplinary actions, and criminal convictions?
2. Is the disclosure of information about malpractice settlements unfair to physicians since many insurance companies make a practice of settling out of court?
3. Would allowing public access to the databank result in the withholding of information concerning physician mistakes and disciplinary actions by hospitals and state medical licensing boards?

Case 4.39 AMA Fights for Control of Price Data

Description

Mr P set up a Web site, *Myhealthscore.com*, to provide consumers with information on health care costs. Prices for medical procedures were estimated using the amounts the federal government reimburses doctors for treatments. Consumers found the information valuable in determining the cost of surgical procedures. One woman, who accessed the site to learn that the government reimbursed physicians $12,000 for a hip replacement, stated, "Doctors freak out when you ask about cost. It's like they want to keep how much they get paid a secret."

Initially, Mr P planned to finance the service by charging doctors for a listing on the Web site, but before he could do so, the AMA ordered him to discontinue listing prices of medical procedures or to pay a significant royalty for doing so. The AMA claimed that it owns the list of codes and descriptions, called Common Procedural Terminology (CPT), that Mr. Pickering used to match payment information with medical procedures. The CPT codes are used by physicians, insurers, and government agencies and were adopted in 1983 by the federal government for use in Medicare, the health insurance program for the elderly and disabled. Federal regulations have also been announced that will make the codes standard for electronic transactions involving physician services.

The AMA has been involved in several lawsuits that have resulted in confidential settlements preserving the AMA's control of the codes. The AMA has also joined with Realtors and providers of legal information to back a bill in Congress that would provide legal protection for databases including the CPT codes. The AMA states that it is opposed to Web sites that encourage consumers to shop for medical care primarily on the basis of price.

The AMA currently has a significant financial stake in controlling the use of CPT codes since it generates millions of dollars annually by licensing the codes to publishers and software distributors who market to physicians. In response to AMA objections, Mr. Pickering discontinued publishing physician price data, and instead, the *MyHealthscore.com* site now offers price data on hospital inpatient services. This information is based on a nonproprietary set of codes available from the federal government.

Source: Carrns A. AMA fights for control over doctor-price data Web sites are providing. *The Wall Street Journal.* August 25, 2000: A1,A5.

Questions for Discussion

1. Should the public have the right to physician price information based on the CPT codes? Why, or why not?
2. Is it unfair to patients who pay for physician services out-of-pocket to withhold price information from them?
3. Is access to price information essential to the success of "defined contribution plans" in which employees are given a fixed amount of money to pay for their own health services?

Protection of Patients' Privacy

Case 4.40: Confidentiality of Patient Data

Description

A hospital is installing a medical information system. Currently physicians are able to access registration and laboratory information on patients. The hospital initially decided that, for reasons of confidentiality, the information that could be accessed on each terminal would be limited to those patients on that terminal's nursing unit. This meant that physicians needed to walk to each nursing unit to access information on their patients in that unit.

After complaints from physicians, the policy was changed, and some physicians were given a password that permitted them to access their patients' records from any location. As more physicians requested the same privilege, the hospital reverted back to a policy by which information on patients could be accessed from any location without a password.

Source: JD Miller, personal communication.

Questions for Discussion

1. Did the initial policy of limiting access to records of patients on a hospital unit unnecessarily restrict the value of the information system to health care providers?
2. Did the initial policy violate the physician's autonomy?
3. Are patients at risk when physicians have limited access to their clinical information?
4. Who is accountable in case a patient is harmed due to the lack of timely access to clinical information? The physician? The hospital?

Case 4.41: Legislated Restrictions on Access to Medical Records for Research

Description

The state of Minnesota has enacted privacy legislation to protect medical records. As of January 1, 1997, health care providers must notify patients in writing that medical records may be released for research purposes and that the patient may object to this use of their health information. Patients must authorize any release of their medical records, and patients whose medical data is used for research purposes must be told how they can contact the investigators.

The policy appears to be predicated upon the assumption that it is inappropriate for anyone to see the medical record except the patient's personal physician. Critics point out that this legislation, by placing restrictions on access to medical data for research purposes, has the potential to negatively affect health research. For example, it may be impossible to obtain specific prior consent from a patient each time a record is used for research, especially in the case of retrospective studies that assess disease trends over long periods of time and long-term outcomes of treatment. In cases in which a patient has died since being treated, consent would have to be obtained from the next of kin. The difficulty in locating living patients or the next of kin of deceased patients would introduce biases into the data that might affect research findings.

Source: Melton LJ. The threat to medical records research. *N Engl J Med.* 1997; 2337:1466–1469.

Questions for Discussion

1. Does the requirement to track the use of records for research impose an unnecessary burden on institutions and individuals engaged in health research?

2. Do these restrictions make it impossible logistically to perform chart reviews and retrospective studies to assess treatment outcomes and trends?
3. Should these confidentiality requirements expire when the patient dies, or should research have to obtain permission to use the patient's medical record from next of kin?
4. How can we balance the conflict between patients' privacy and the need for access to medical data for research purposes?

Hackers

Case 4.42: Beware: Hackers at Play

Description

A group of high-school students broke into more than 60 business and government computer systems in the United States and Canada. One of these systems was that of the Memorial Sloan-Kettering Cancer Center in New York. All that was necessary to penetrate the computers at Sloan-Kettering, Los Alamos, and elsewhere was a home computer, a modem, and a modicum of computer literacy. One hacker told investigators that the group gained access to Sloan-Kettering's computer system using the user name "test" and the password "test." They gained privileged user's status through a help menu with the command "set process/PRIV=all."

The youths worked mostly at night using personal computers. They would randomly try numbers until they reached a computer that answered back. The way the computer system responded indicated what type of system it was. Since each system has its own format for allowing user access, knowing the type of system allowed the hackers to narrow the range of possible passwords. In some instances, they were able to obtain passwords from bulletin boards. Frequently, the computer system would respond to incorrect password with the message "unauthorized access" and then allow them to keep trying other passwords.

Source: Marbach WD. Beware hackers at play. *Newsweek.* September 5, 1983: 42–46.

Questions for Discussion

1. Do you agree with the assertion that since the youths caused no harm they should not be prosecuted? Why? Why not?
2. Will it be difficult to prosecute them without proof of criminal intent?
3. Was the cancer center lax in protecting its computer system from hackers? What should it have done differently?

Case 4.43: Computer Engineer Sentenced

Description

A computer engineer for a nationwide hospital software firm was sentenced to

six months in prison for hacking files at an Illinois medical center. "I don't know why I did it," the engineer told US District Court. The crime led to fines and restitution of $31,425.
Source: *Computers & Medicine*. May 1999; 28:16.

Questions for Discussion

1. How can medical center information systems be protected against hackers?
2. How can one balance the need for security against the need for access to medical information by providers from a number of different locations?

Case 4.44: University Tightens Computer Security

Description

A university is tightening its computer security after hackers broke into a computer at the medical school and secretly used it to generate a flood of e-mail advertisements. Efforts by the university to cope with the break-in have caused balky and intermittent e-mail service for 7 months for hundreds of staff members. At least once, e-mail service throughout the system shut down for 2 days. University officials didn't detect the break-in until at least a couple of weeks later, when someone forwarded an advertisement sent by the computer.

A university spokesperson said that no file information was improperly accessed. Instead the hackers merely used the system to generate e-mail promoting other Web sites. The university announced that $150,000 would be spent to install new equipment to restore the e-mail system. A number of security measures were being upgraded to prevent the computer system from being broken into in the future.
Source: Birch D. Hopkins tightens computer security. *The Baltimore Sun*. May 29, 1999: 1B–2B.

Questions for Discussion

1. Are university medical center information systems especially vulnerable to hackers? Why, or why not?
2. Is the medical center accountable for any harm that is caused by unauthorized entry into patient records?

Privacy Online

Case 4.45: Uncovering Physicians' Online Identities

Description

About 50 physicians are defendants in a libel complaint filed by PhyCor, a physician practice management company. The identities of physicians who had posted criticisms of PhyCor on a Yahoo! message board were revealed in response to subpoenas issued as a result of the company's complaint. PhyCor alleges that the physicians falsely defamed the company and published legally prohibited information about the company. A number of physicians have been informed by their Internet service providers that their identities have been revealed in response to subpoenas issued as a result of the complaint by PhyCor.

John Doe libel lawsuits are being used by more companies to combat the rise in anonymous criticisms of the company or of any of its administrative officers on electronic message boards. Many of the messages concerning PhyCor criticized the company's business model. PhyCor purchases the assets of clinics, pays shareholder physicians a fee, and then charges the clinics a management fee in exchange for managed care contracts. Physicians affiliated with PhyCor are also particularly unhappy about the plunge in the value of the company's stock.

Source: Tokarski C. PhyCor moves to uncover physicians' online identities. *Am Med News.* 1999; 42:1,38.

Questions for Discussion

1. Were the messages posted on the bulletin boards free speech? Is the medium an appropriate one for health professionals to voice grievances?
2. Did the Internet service provider violate the confidentiality of the physicians when it disclosed their identities to PhyCor?
3. Will it be more difficult for PhyCor to contract with other clinics and physicians as a result of the negative publicity surrounding the legal proceedings?

Case 4.46: CEO Sues Online Critics

Description

Ms L, a former HealthSouth employee, and others who posted complaints on a Yahoo! message board devoted to the company, are being sued for defamation. HealthSouth provides rehabilitation services. The CEO, in response to criticisms posted online, asked for the records of each of the 300 anonymous individuals who had posted messages on the Web site. When Yahoo! officials refused to provide all but 20 records, the CEO hired a detective to identify the authors of the messages, sued many of the authors, and in a few instances, pressed criminal charges.

Company employees watched the bulletin board for reports about the company since its stock value has begun to slip. Some of these messages advised stockholders to sell their stock and claimed that the company was using a pyramid scheme to inflate its earnings. Other messages poked fun at the CEO and his management style or were more personal, for example, discussing the CEO's family relations.

Source: Moss M. CEO exposes, sues anonymous online critics. *The Wall Street Journal.* July 7, 1999: B1,B4.

Questions for Discussion

1. Will legal action and the use of private investigators to uncover the identity of persons who post messages on electronic bulleting boards limit the freedom of expression of users?
2. Are companies using the threat to sue critics to intimidate them?
3. Is the use of bulletin boards to criticize or comment about the company merely an expression of free speech?

Note: Parts of this chapter are taken from Anderson JG, Brann M. Security of Medical information: The threat from within. MD Computing. *2000; 17:15–17; and Anderson JG. Security of the distributed electronic patient record: A case based approach to identifying policy issues.* Int J Med Inf *2000;60:111–118, with permission.*

References

1. Rindfleisch TC. Privacy information, technology, and health care. *Commun ACM.* 1997;40:93–100.
2. National Research Council, Committee on Maintaining Privacy and Security in Health Care Applications of the National Information Infrastructure. *For the record: protecting electronic health information.* Washington, DC: National Academy Press; 1997.
3. Gallo AC, Lee VJ. *Health care information technology: keeping health care wired,* Research Report. Baltimore: Alex Brown; 1998.
4. Dick RS, Steen EB, Detmer DE, eds. *The computer-based patient record,* Rev. ed. Washington, DC: National Academy Press; 1997.
5. Moran DW. Health information policy: On preparing for the next war. *Health Affairs.* 1998;17:9–22.
6. Simpson RL. Security threats are usually an inside job. *Nurs Manage.* December 27, 1996:43.
7. Kleinke, JD. *Bleeding edge: the business of health care in the next century.* Gaithersburg, Md: Aspen 1998.
8. Tang PC, Hammond WE. A progress report on computer-based patient records in the United States. In: Dick RS, Steen EB, Detmer DE, eds. *The computer-based patient record: an essential technology for healthcare,* Rev. ed. Washington, DC: National Academy Press; 1997:1–20.
9. van Bemmel JH, van Ginneken AM, van der Lei J. A progress report on computer-based patient records in Europe. In: Dick RS, Steen EB, Detmer DE, eds. *The computer-based patient record: an essential technology for healthcare,* Rev. ed. Washington, DC: National Academy Press; 1997:21–43.
10. I/T sales to soar next five years. *Health Manage Technol.* December 1995: 10.
11. Kleinke JD. Release 0.0: Clinical information technology in the real world. *Health Affairs.* 1998;17:23–38.
12. Anderson JG. Clearing the way for physician use of clinical information systems. *Commun ACM.* 1998;40:83–90.
13. Goldman J. Protecting privacy to improve health care. *Health Affairs* 1998;17:47–60.

14. Horovitz B. 80% fear loss of privacy to computers. *USA Today*. October 31, 1995:A1.
15. Laidman J, Woods M. Sex doctor's patient files show up on the Web. *Pittsburgh Post-Gazette*. March 28, 1999. Available at http://www.post-gazette.com/headlines/ 19990328doclist2.asp. Accessed October 1, 2001.
16. Bass, A. HMO puts confidential records on-line. *Boston Globe*. March 7,1995: 1.
17. Weinstein L. Confidential patient data accidentally released to the Web. Privacy Forum, February 20, 1999. Available at: http://www.vortex.com/privacy/priv.0.8.04. Accessed October 10, 2001.
18. Goldman J, Mulligan D. *Privacy and health information systems: a guide to protecting patient confidentiality*. Washington DC: Center for Democracy and Technology; 1996.
19. Siegler M. Confidentiality in medicine—a decrepit concept. *N Engl J Med*. 1982;307:158–21.
20. Davis R. Online medical records raise privacy fears. *USA Today*. March 22, 1995: 1A.
21. Upton J. U-M medical records end up on Web. *The Detroit News*, February 12, 1999: 1A.
22. Antommaria AHM. Private correspondence. June 16, 1999.
23. Weingarten J. Can confidential patient information be kept private in high-tech medicine? *MD Computing*. 1992;9:79–82.
24. Shalala DE. Health care information and privacy. *Health Matrix J Law Med*. 1998;8:223–232.
25. Rogers L, Leppard D. For sale: Your secret medical records for 150 pounds. *London Sunday Times*. November 26, 1995:1–2.
26. Hospital clerk's child allegedly told patients that they had AIDS. *Washington Post*. March 1, 1995:A17.
27. Medical Information Bureau. *The Consumer's MIB Fact Sheet*. Westwood, Mass: Medical Information Bureau, 1991.
28. Geller LN, Alper JS, Billings PR, Barash CI, Beckwith J, Natowicz MR. Individual, family, and societal dimensions of genetic discrimination: A case study analysis, *Sci Eng Ethics*. 1996;2:71–88.
29. Pendrak RF, Ericson RP. Information technologies need to protect patient confidentiality. *Healthcare Financial Manage*. October 1998:1–3.
30. Congressional Office of Technology Assessment. *Protecting privacy in computerized medical information*. Washington, DC: US Government Printing Office, 1993. Publication OTA-TCT-576.
31. Foley J. Data dilemma. *Information Week*. June 10, 1996:14–16.
32. Bernstein N. Lives on file: The erosion of privacy—a special report. *The New York Times*. June 12, 1997:A1.
33. Harrow Jr. RO. Prescription sales privacy fears; CVS, Giant share customer records with drug marketing firm. *Washington Post*. February 15, 1998: A1.
34. PRNewsire: Minnesota takes the lead on agreement to protect 41 million Americans, October 25, 1999. Available at: http://www.epic.org/privacy/medical/merck.txt. Accessed October 10, 2001.
35. Davis R. On-line medical records raise privacy fears. *US Today*. March 22, 1995, A1.
36. Petersen A. A privacy firestorm at DoubleClick. *The Wall Street Journal*. February 23, 2000:B1,4.

37. Tierney WM, Murray MD, Gaskins DL, Zhou XH. Using computer-based medical records to predict mortality risk for inner-city patients with reactive airways disease. *J Am Med Inform Assoc*. 1997;4:313–321.

38. Tierney WM, Takesue BY, Vargo DL, Zhou XH. Using electronic medical records to predict mortality in primary care patients with heart disease: Prognostic power and pathophysiologic implications. *J Gen Intern Med*. 1996;11:83–91.

39. Schwartz PM, Reidenberg JR. *Data privacy law: a study of United States data protection*. Charlottesville, Va: Michie Law Publishers, 1996.

40. Thomas Legislative Information on the Internet. Available at: http://thomas.loc.gov. Accessed October 9, 2001.

41. Health Insurance Portability and Accountability Act. Available at: http://aspe.hhs.gov/admnsimp/. Accessed October 10, 2001.

42. Computer Science and Telecommunications Board, National Research Council. *For the record: protecting electronic health information*. Washington, DC: National Academy Press; 1997.

43. Computer Science and Telecommunications Board, National Research Council. *Networking health—prescriptions for the Internet*. Washington, DC: National Academy Press, 2000.

44. Goldman J. Protecting privacy to improve health care. *Health Affairs*. 1998;17:47–60.

45. US Department of Health and Human Services. *Protecting human research subjects: institutional review board guide book*. Washington, DC: US Government Printing Office, 1993.

46. Szolovits P, Kohane I. Against simple universal health care identifiers. *J Am Med Inf Assoc*. 1994;1:316–319.

47. Schwartz PM. European data protection law and restrictions on international data flows. *Iowa Law Review*. 1995;80:471–496.

48. Donaldson MS, Lohr KN, eds. *Health data in the information age: use, disclosure and privacy*. Washington, DC: National Academy Press; 1994.

49. Office of Technology Assessment. *Bringing health care online: the role of information technologies*. Washington, DC: U.S. Government Printing Office; 1995.

50. Moran DW. Health information policy: On preparing for the next war. *Health Affairs*. 1998;17:9–22.

Further Readings

Advisory Committee on Automated Personal Data Systems. *Records, computers and the rights of citizens*. Washington, DC: Department of Health, Education, and Welfare; 1973.

Allen A. *Uneasy access*. Totowa, NJ: Rowman and Littlefield Publishers; 1988.

Alpert SA. Health care information: access, confidentiality and good practice. In: Goodman KW, ed. *Ethics, Computing and Medicine*. New York, NY: Cambridge University Press; 1998:75–101.

Alpert SA. Smart cards, smarter policy: Medical records, privacy and health care reform. *Hastings Center Report*. 1993;23;13–23.

Annas GJ. Privacy rules for DNA databanks: protecting coded 'future diaries'. *JAMA*. 1993;270:2346–2350.

Anthony J. Who's reading your medical records? *Am Health* November 1993:54–58.

Bakker AR. Security in medical information systems. In: *Yearbook of medical informatics '93.* Stuttgart Germany: Shattauer; 1993:52–60.

Barber B. Current issues in data protection. *Med Inf.* 1989;14:207–209.

Barber B, Treacher A, Louwerse CP, eds. *Toward security in medical telematics: legal and technical aspects.* Amsterdam: IOS Press; 1996.

Barrows RC, Jr., Clayton PD. Privacy, confidentiality, and electronic medical records. *J Am Med Inf.* 1996;3:139–149.

Baskersville R. *Designing information systems security.* Chichester, UK: Wiley & Sons; 1988.

Bennett C. Can on-line health care seriously damage your privacy? *Chicago Tribune.* October 28, 1999:Sect 1, 15.

Bennett CJ. *Data protection and public policy in Europe and United States.* Ithaca, NY: Cornell University Press; 1992.

Biskup J. Protection of privacy and confidentiality in medical information systems: Problems and guidelines. In: Spooner DL, Landweher C, eds. *Database security.* Amsterdam: Elsevier Science Publishers; 1990.

Bleumer G. Security for decentralized health information systems. *Int J Biomed Comput.* 1994;35(Suppl 1):139–145.

Bollas C, Sundelson D. *The new informants: the betrayal of vonfidentiality in psychoanalysis and psychotherapy.* Northvale, NJ: Jason Aronson Inc, 1995.

Brannigan VM. Patient privacy: A consumer protection approach. *J Med Systems.* 1984;8:501–505.

Brannigan VM. Protecting the privacy of patient information in clinical networks: Regulatory effectiveness analysis. In: Parsons DF, Fleischer CN, Greene RA, eds. *Extended clinical consulting by hospital computer networks.* New York, NY: Annals of the New York Academy of Sciences; 1992.

Brannigan VM, Beier B. Standards for privacy in medical information systems: A technico-legal revolution. *Datenschutz and Datensicherung.* September 1991.

Chadwick DW, Crook PJ, Young AJ, McDowell DM, Dornan TL, New JP. Using the Internet to access confidential patient records: a case study. *BMJ.* 2000;321:612–614.

Commission of the European Communities DG XIII/F AIM. *Data protection and confidentiality in health informatics.* Washington, DC: IOS Press; 1991.

Computer Science and Telecommunications Board, National Research Council. *For the record: protecting electronic health information.* Washington, DC: National Academy Press; 1997.

Computer Science and Telecommunications Board, National Research Council. *Networking health prescriptions for the Internet.* Washington, DC: National Academy Press; 2000.

Computers and privacy: how the government obtains, verifies, uses and protects personal data. Washington, DC: General Accounting Office; August 1990.

Dick RS, Steen EB, Detmer DE, eds. *The computer-based patient record: an essential technology for healthcare* Rev. ed. Washington, DC: National Academy Press; 1997.

Doctors' and pharmacies' files are gathered and mined for use by drug makers. *The Wall Street Journal.* February 27, 1992: A1.

Donaldson MS, Lohr KN, eds. *Health data in the information age: use, disclosure and privacy.* Washington, DC: National Academy Press; 1994.

Feehan KP. Legal access to patient health records/Protection of quality assurance activities. *Health Law Can.* 1991;12:3.

Flaherty DH. *Protecting privacy in surveillance societies: The Federal Republic of Germany, Sweden, France, Canada, and the United States.* Chapel Hill, NC: University of North Carolina Press; 1989.

Flaherty DH. Privacy, confidentiality, and the use of Canadian health information for research and statistics. *Can Public Admin.* 1992;35:80.

Furnell SM, Gaunt PN, Pangalos G, Sanders PW, Warren MJ. A generic methodology of health care data security. *Med Inf.* 1995;19:229–246.

Gaunt N, Roger-France F. Security of the electronic health care record—professional and ethical implications. In: Barber, B et al. eds. *Towards security in medical telematics.* Amsterdam: IOS Press; 1996.

Gavison R. Privacy and the limits of the law. In: Schoeman FD, ed. *Philosophical dimensions of privacy: an anthology.* Cambridge, UK: Cambridge University Press, 1984.

Gellman RM. Prescribing privacy: the uncertain role of the physician in the protection of patient privacy. *NC Law Rev.* 1984;62:258.

Goldman J, Mulligan D. *Privacy and health information systems: a guide to protecting patient confidentiality.* Washington, DC: Center for Democracy and Technology, 1996.

Gostin LO. Health information privacy. *Cornell Law Rev.* 1995;80:101–184.

Gostin LO, Turek-Brezina J, Powers M, Kozloff R, Faden R, Steinauer DD. Privacy and security of personal information in a new health care system. *JAMA.* 1993;270:2487–2493.

Griesser G, Bakker A, Danielsson J, Hirel JC, Kenny DJ, Schneider W, et al. *Data protection in health information systems: considerations and guidelines.* Amsterdam: North Holland; 1980.

Gritzalis D, Katsikas S, Keklikoglou J, Tomaras A. Data security in medical information systems: technical aspects of a proposed legislation. *Med Inf.* 1991;16:371–383.

Hamilton DP. Freedom software lets you get some privacy while surfing the Web. *The Wall Street Journal.* August 10, 2000: B1.

Hammond WE. Security, privacy and confidentiality: A perspective. *J Health Infor Manage Res.* 1992;1:1–8.

Hendricks E, Hayden T, Novik JD. *Your right to privacy: a basic guide to legal rights in an information society*, 2nd ed. Carbondale, Ill: Southern Illinois University Press; 1990.

Kolata G. When patients' records are commodities for sale. *The New York Times.* November 15, 1995: B1.

Kluge EH. Advanced patient records: Some ethical and legal consideration touching medical information space. *Methods Inf Med.* 1993;32:95–103.

Lawrence LM. Safeguarding the confidentiality of automated medical information. *J Quality Improvement.* 1994;20:639–645.

Linowes DF. *Privacy in America.* Urbana, IL: University of Illinois Press; 1989.

Medical Information Bureau. *The consumer's MIB fact sheet.* Westwood, Mass: Medical Information Bureau; 1991.

Medical Information Bureau. *MIB, Inc.: a consumer's guide.* Westwood, Mass: Medical Information Bureau; 1990.

Moehr JR. Privacy and security requirements of distributed computer based patient records. *Int J Biomed Comput.* 1994;35(Suppl 1):57–64.

Murphy G. System and data protection. In: Ball MJ, Collin MF, eds. *Aspects of the computer-based patient record.* New York: Springer-Verlag; 1992.

Oates R. Confidentiality and privacy from the physician perspective. In: *Compendium of the First Annual Confidentiality Symposium of the American Health Information Management Association.* July 15, 1992, Washington, DC; 138–143.

Pfleeger SL. A framework of security requirements. *Computers and Security.* 1991;10:515–523.

Privacy act: federal agencies' implementation can be improved. Washington, DC: General Accounting Office; August 1986.

Rienhoff O. Digital archives and communication highways in health care require a second look at the legal framework of the seventies. *Int J Biomed Comput.* 1994;35(Suppl 1):13–19.

Roach WH Jr, Chernoff SN, Esley CL, eds. *Medical records and law.* Rockville, Md: Aspen Systems Corp; 1985.

Robinson EN Jr. The computerized patient record: privacy and security. *MD Comput.* 1994;11:69–73.

Rothfeder J. *Privacy for sale.* New York: Simon & Schuster; 1992.

Safran C, Rind D, Citreon M, Bakker AR, Slack WV, Bleich HL. Protection of confidentiality in the computer-based patient record. *MD Comput.* 1995;12:187–192.

Schwartz PM, Reidenberg JR. *Data privacy law: a study of United States data protection.* Charlottesville, VA: Michie Law Publishers; 1996.

Shea S. Security versus access: Trade-offs are only part of the story. *J Med Inf Assoc.* 1994;1:314–315.

Skolnick AA. Protecting privacy of computerized patient information may lie in the cards. *JAMA.* 1994;272:187–189.

US Congress, Office of Technology Assessment. *Automated medical records: leadership needed to expedite standards of development.* Washington, DC: US Government Printing Office; 1993. Publication GAO/IMTEC-93-17.

US Congress, Office of Technology Assessment. *Protecting privacy in computerized medical information.* Washington, DC: US Government Printing Office; 1993. Publication OTA-TCT-576.

US Congress, Office of Technology Assessment. *Electronic record systems and individual privacy.* Washington, DC: US Government Printing Office; 1986. Publication OTA-CIT-296.

US Congress, Office of Technology Assessment. *Defending secrets, sharing data: new locks and keys for electronic information.* Washington, DC: US Government Printing Office; 1987. Publication OTA-CIT-310.

US Congress, Office of Technology Assessment. *Medical monitoring and screening in the work place: results of a survey—background paper.* Washington, DC: US Government Printing Office; 1991.

US Congress, Office of Technology Assessment. *Bringing health care online: the role of information technologies,* Washington, DC: US Government Printing Office; 1995. Publication OTA-ITC-0036.

US Privacy Protection Study Committee. *Personal privacy in an information society.* Washington, DC: US Government Printing Office; 1977.

Van der Leer OF. The use of personal data for medical research: How to deal with new European privacy standards. *Int J Biomed Comput.* 1994;35(Suppl):87–95.

Wald JS, Law M, Meade T, Miller G, Alberman E, Dickinson J. Use of personal medical records for research purposes. *BMJ.* 1994;309:1422–1424.

Weingarten J. Can confidentiality of information be kept private in high-tech medicine? *MD Comput.* 1992;9:79–82.

Westin A. *Computers, health records, and citizen rights.* Washington, DC: US Government Printing Office; 1976.

Ziporyn T. Hippocrates meets the data banks: patient privacy in the computer age. *JAMA.* 1984;252:317–319.

5

The Challenge of Bioinformatics

Reprinted with permission from Creators Syndicate.

One could make a very strong argument that information technology and genetics are the sciences that will have the greatest effect on twenty-first century health care. Although it is already clear that informatics is profoundly reshaping the health professions, we are only beginning to come to terms with the extraordinary risks and potential benefits of progress in the human genome sciences. The intersection of health informatics and genomics, along with the consequent ethical and social issues, confront us with one of the greatest intellectual and practical challenges in the history of science.

Bioinformatics, the use of information technology to acquire, store, manage, share, analyze, represent, and transmit genetic data, has blossomed in the past several years. The term is most often used by scientists who sequence and otherwise analyze the genomes of humans and other species with computer technology. If we like, we can stipulate that bioinformatics includes applications as pedestrian as using a personal computer to store the results of genetic tests (Patient X has the BRCA1 gene) as well as the use of intelligent machines to link physiological traits with a database to diagnose genetic maladies, predict clinical correlations, conduct research, and so forth.

The future of genomics will likely be built upon vast amounts of computer-based information, from data acquisition via "gene chips" to diagnosis using genetic decision support systems to research based on data mining of very large databases and warehouses. Social and ethical issues raised by the genomic sciences will stimulate clinicians and scientists at least as much as those raised by clinical informatics. The broad and rapid growth of bioinformatics presents exciting opportunities and challenges not only for clinicians and scientists but also for society, individuals, policy-making bodies, and governments.

Although bioinformatics raises many issues for research using human subjects, we will confine ourselves here to more clinical concerns and group these concerns under the label "clinical bioinformatics." (Note, though, that once human genetic information is stored on a computer, it is much easier to study; in some cases the distinction between clinical and research issues will narrow dramatically.) Let us organize the ethical and social issues raised by clinical bioinformatics into the following categories: (i) accuracy and error, (ii) appropriate uses and users of digitized genetic information, and (iii) privacy and confidentiality.

Accuracy and Error

Experience with health informatics has taught us that accuracy and error avoidance raise ethical issues that are often related to evolving standards of care. If there are emerging or established standards for database management, for instance, then a system that relies on a database will be more or less useful, reliable, and safe, depending upon whether or not the database is appropriately maintained, tested, augmented, and so on. The reason to link error and ethics is that errors, however unintentional, can produce harm. Determining whether a harm constitutes a wrong is one of the main challenges of ethics. Our specific challenge here is to nurture the growth of an exciting new science while simultaneously ensuring that patients are not harmed or wronged.

Several current and future issues related to the accuracy of bioinformatics systems include:

Risks to persons. To the extent that we can expect more and more frequent computer-aided discoveries of the genetic loci of human diseases, errors can pose or increase risks to public health and even the well-being of individuals. Patients may also be at risk when computers are used to predict the expression of future genetic maladies. The risks may be psychological and will likely vary depending on whether there is a treatment or cure for a given malady. The role of genetic counselors will loom large here.

Recanted linkage studies. Preliminary or unreplicated linkage studies are sometimes recanted or re-evaluated. Erroneous linkage analyses can throw colleagues off the track and, perhaps more importantly, cause unnecessary psychological trauma for individuals who fear they may be affected. In the case of purported linkages that correlate with race or ethnicity, there is the added risk of producing social stigma, perhaps especially in the case of neurogenetics and psychiatric genetics.

Meta-analysis. It is exciting to observe the emergence of meta-analysis in genomics. Meta-analysis involves the aggregation and reanalysis of the results of previous studies by statistical software with the aim of achieving statistical significance or adequate sample sizes. This technique raises ethical issues partly by virtue of doubts about the quality of included data and the validity of inferences based on diversity of data. These doubts are important when meta-analytic results are applied to patient care.

Decision support. Although diagnostic and decision support systems are well known to raise ethical issues in clinical medicine, there is, as yet, no critical analysis of decision support for genetic diagnoses in which, for example, clinical information, photographic material, pedigree, and gene localization data are analyzed by computers. The growth of genomic databases and the increasing availability of genetic information at the clinical level suggest that decision support systems are a ripe source for ethical and social inquiry.

Appropriate Uses and Users

Questions concerning who should use clinical information systems and in what contexts have been shown to raise interesting and important ethical issues. We should expect that processing genetic data will elicit related concerns and pose new problems.

For instance, suppose a physician or nurse begins to include genetic data in patient charts, uses those data to predict the likelihood of clinical manifestations and correlations, and employs those analyses to refer patients to genetic counselors. The first question is basic: Was this novel use undertaken with the patient's consent? Because genetic information can frighten or alarm patients in ways that other health and medical data do not, we need to ask whether the patients knew that genetic data was being gathered and stored for clinical purposes. In the absence of a treatment or cure for a particular genetic malady, it is not unreasonable for a patient to prefer not to know a genetic diagnosis or prognosis. Consent seems to be a crucial gate through which the physician or nurse must pass before using these data "for the patient's sake." The weight of valid or informed consent seems greater here—that is, in the area of genetics—than for more familiar kinds of clinical decision support.

To raise another concern, suppose that individuals' genetic data were being collected by governments, managed-care organizations, or other third-party payers with the goal of shaping or adjusting risk pools or coverage eligibility. The difference between evidence-based actuarial calculations and discrimination can be very slight, indeed. To the extent that computers are used for these tasks, it will be essential for individuals, institutions, and society to decide on ethically optimized strategies for clinical applications of bioinformatics.

Who should use a genetic diagnostic or prognostic system? For example, does the possibility that bioinformatics tools could be used in problematic ways to determine health benefits imply that bioinformatics tools should never be used by certain entities? Should certain types of use be prohibited? And how can such a prohibition be enforced? Consider that individual physicians, nurses, genetic counselors, or psychologists might use computer systems not only to improve patient care but also for less worthy purposes—discrimination against people with genetic meladies, say. Does it follow that certain users—in addition to uses—might be problematic?

One way to approach the question is to ask whether the user is trained to employ a computer in a task not normally within his or her competence. For instance, if a physician or nurse does not normally render genetic diagnoses, it is unwise to suppose that she or he acquires competence via the machine.

In fact, it is more than unwise; it is a patent mistake. Computers can improve our skills at many tasks but rarely, if ever, give us new professional skills or abilities. Therefore, the appropriate role for a genetic decision-support system, for instance, will be to assist adequately trained professionals, not to replace them or to bring them "up to speed" in domains in which they lack basic skills.

This point must be clearly understood: Computers can be outstanding educational tools in bioinformatics as elsewhere, but there is a difference between acquiring a skill and presuming its existence. We have learned from "ordinary" clinical computing that humans practice medicine and nursing but computers do not. This is a lesson well worth applying to bioinformatics.

Privacy and Confidentiality

The electronic storage of genetic information replicates a tension already familiar in health informatics: the tension between (i) the need for appropriate or authorized access to personal information, and (ii) the need to prevent inappropriate or unauthorized access. Striking a balance between these two imperatives is an exciting but sometimes vexing challenge.

Privacy and confidentiality are potentially threatened when an individual's genetic data are maintained or transmitted using computers. The threats include bias and discrimination, personal stigma (as opposed to population or subgroup stigma), psychological stress, and tensions within families, among other risks. We do not yet know whether the inclusion of genetic data adds to or alters the difficulties posed by expectations of privacy and confidentiality of electronic patient record.

Specifically, our objective is to determine if and in what way bioinformatics raises ethical issues to distinct from traditional approaches to ethics and genetics and to adapt existing conceptual and pedagogic tools to our findings or provide new ones. The key means by which we plan to meet these objectives are the development of ethically optimized guidelines (for organizations that maintain databases, for institutional review boards, etc.) and model curricula in ethics and bioinformatics (for students and professionals).

Striking a Balance

The need for organizational policies, best-practice standards, and/or guidelines is widespread in the human sciences. Because the thrust of the proposed research is at the intersection of three vast areas of inquiry and practice— genetics, computing, and ethics—the challenge we face is extraordinary. Guidelines and standards often fail because they are either so broad or simplistic that they cannot adequately guide behavior, or so specific or detailed that they are too inflexible to be useful in diverse and unexpected cases. There is, therefore, a need to strike a balance between these two shortcomings. In striking that balance, the successful completion of this project would provide a very useful tool for organizations.

Educational materials for research ethics curricula usually overlook issues in bioinformatics. If we are correct in anticipating that the future of genetic research will be inextricably linked to information-processing technologies, then this oversight is, or will be, quite serious.

Case Studies

Issues in Bioinformatics

Case 5.1: Computational Pharmacogenomics

Description

The announcement in the summer of 2000 that scientists had completed a rough draft of the sequence of the human genome has set off a race that might prove as frenzied, fraught, and controversial as the competition to sequence the genome in the first place. This new race is to use this information to develop drugs customized to the genetic make-up of individual patients, to better predict response to drugs, and to better automate pharmaceutical discovery.

Microarrays and gene chips are near the center of this new science and industry. Microarrays are, in a way, a combination of the computer and the wet lab that can, in principle, produce rapid and comprehensive analyses of individuals' genetic structure. The machine identifies mutations, polymorphisms, and other features in a patient's cells.

But to get to that point, scientists need to create DNA banks or reference libraries using tissue donations from many individuals to assess the breadth and depth of human gene function and variety. Increasingly, the information contained in our genes is digitized, and stored, analyzed, and shared using computers. Consider, then, The DNA Sciences Gene Trust Project, an effort by DNA Sciences Inc. As explained it one time on the company's Web site (http://www.dna.com):

> In creating The Gene Trust, our goal is to establish a huge database of information about people—physical characteristics, health histories, responses to treatments, etc. Based on the genetic knowledge this will give us, we can drastically speed up the rate of medical advances.

Among the reasons for setting such a goal is the firm's intent to

> Develop intellectual property regarding genetic profiles and their link to [. . .] tests for screening common disease and to help physicians and individuals determine the right therapeutic treatment for these common diseases [and] Perform the screening/personalized medicine tests and/or license this intellectual property to third parties so that they can commercialize these tests.

The result is a scenario in which a patient is asked to provide a blood sample for microarray analysis, digitization, and storage. After some point, the biological sample will be discarded, but the genetic information it held will be retained on one or more computers.

Sources: DNA Sciences Inc, available at: http://www.dna.com; Georgetown University's Cytochrome P450 Drug Interaction Table, available at: http://www.drug-interactions.com.; and United States Department of Energy and National Institutes of Health Human Genome Project's pharmacogenomics site, http://www.ornl.gov/hgmis/medicine/pharma.html. All accessed October 8, 2001.

Questions for Discussion

1. Challenges related to banked tissue research are increasingly well known. Do digitization and electronic storage of genetic information raise new issues? Many hospitals and medical centers already store vast amounts of human tissue, blood and, other material. These repositories are increasingly a source of research material. Now imagine that all the genetic information in those freezers and storage cabinets is online, ready for database analysis, e-mail and marketing.

2. Drug discovery and refinement will increasingly rely on blood and other

biological contributions from people unlikely to benefit in the short term. What duties do pharmaceutical and other companies have to people who are the sources of genetic material/information—especially when that information can be electronically immortalized?

3. Some bioinformatics Web sites simultaneously note the scientific benefits of drug research and seek contributions of biological material. Suppose it were implied that blood contributions are altruistic. (The Gene Trust Project cited here suggested that blood sources are being offered "Nothing less than a chance to participate in history.") How vigorously should computational genomics Web sites be scrutinized by institutional review boards? (Compare, also, the discussion and cases in Chapter 6.)

Case 5.2: Genetic Information in Centralized Databases

Description

Individual genetic information is increasingly stored in public, private, and government health databases. The databases are or could be used for clinical practice, epidemiologic research, pharmaceutical investigations and other purposes. The World Medical Association (WMA) is drafting guidelines for the use of genetic information in such databases.

According to WMA Chair Anders Milton, "The public is rightly concerned about whether their right to privacy and confidentiality is threatened by these databases and whether information about them as individuals could be misused. Centralized health databases can make a tremendous contribution to the improvement of health. But the public's right to privacy and consent are essential to the trust and integrity of the patient/physician relationship, and we must ensure that these rights are properly protected. Any guidelines must address the issues of privacy, consent, individual access, and accountability."

- Source: World Medical Association. WMA to draw up health database guidelines. May 8, 2000, Available at: http://www.wma.net/e/press/00_11.html. Accessed October 9, 2001.

Questions for Discussion

1. For years, genetic information has been included in electronic databases without clear regulation or rules. If guidelines are formulated, should they apply to information collected retrospectively, prospectively, or both? Might it ever be too late for guidelines?

2. To what extent can informed or valid consent requirements be loosened if genetic information is anonymous or not linked to identifiable persons? How should the problem of racial or ethnic stigma be addressed in any guidelines?

3. Does—or how does—the purpose of a database have ethical consequences for its use? That is, does it matter if a database is owned by a (i) for-profit corporation, (ii) public health organization, (iii) government?

Case 5.3: Computational Genomics and Intellectual Property

Description

In the late 1980s, the International Union of Crystallography (IUCr) decided on a policy that "permits depositors of 3-dimensional structures of biological macromolecules to ask the PDB [Protein Data Bank, a Brookhaven National Laboratory data base of images and structural data] to delay release of their coordinates for up to one year following publication of their article in an IUCr journal".

The policy is therefore one that requires the concurrence of leading journals. Those which support the embargo or on-hold policy say that the delay allows those who acquired the data both to establish their priority via publication and to use the data for additional work before others who have not invested comparable time or resources. Those who oppose the embargo argue that it impedes scientific progress and violates independent scientific standards for the free flow of information. Many journals are re-evaluating or even changing their policy.

Source: Sussman JL. What's new at the PDB. *PDB* (Newsletter of the Brookhaven National Laboratory) 1998;84: 1.

Questions for Discussion

1. Should computer databases of genetic and related information be always open and as complete as possible, or should the scientists who gathered or organized the information be allowed a "head start" to make the most of their research?
2. Who should be responsible for setting and perhaps even policing standards in this area: The government, perhaps through research funding agencies? Database administrators and overseers? Scientific societies? Journal editors? Consortia of some or all of the above?
3. What would constitute an ethically optimized stance regarding attribution, citation and "authorship" of online genetic data?

Case 5.4: Web-Mediated Paternity Testing

Description

Paternity testing has always raised difficult questions. Now, though, Web sites offer a chance to test a child's paternity without the consent or knowledge of either (putative) parent or of the child (whatever the child's age) using a hair sample from, as one site terms it, the "alleged father."

The companies offer genetic analysis of hair or buccal (cheek) mucosa samples to determine paternity. A man might therefore complete an online form, submit a sample of his and a child's hair and, in a few days, learn via e-mail if he is the father. A woman, unsure of which of several potential candidates is the father of her child, might obtain a hair sample or samples and submit them along with her child's samples. Should anyone have questions about the process, one firm suggests, "For instant answers or advice, chat to a DNA expert online!"

A British Department of Health spokesperson was quoted in one report is saying of one vendor, "There is nothing illegal about the Web site. We are aware of concerns raised by advances in DNA testing, and we are in the process of drawing up a voluntary code of practice on the way companies work."

Sources: Ananova Ltd. Website launches DNA paternity test. Aug. 29, 2000, http://www.ananova.com/news/story/health_uk-dna-internet-science-technology_949444.html. See also http://www.dnanow.com, http://www.genetestlabs.com, and http://www.dnatesting.com. All accessed October 9, 2001.

Questions for Discussion

1. The standard of care in genetic counseling generally requires pre-test and post-test counseling for individuals and couples. Paternity is among the issues raised during such sessions, and couples or individuals are often warned that they might acquire information that could significantly alter or damage relationships. Web-based paternity testing diminishes or eliminates counseling and/or sharing of information and provides these services via e-mail. Should Web-based testing be required to hew to standards employed elsewhere in genetic testing? How so, given that firms providing these services operate across international boundaries?
2. What kinds of caveats or disclaimers are appropriate for such Web-based services?
3. In the absence of laws that might regulate such Web-based testing, could voluntary policies or guidelines have an adequate effect?

Case 5.5: Errors in Genetic Databases

Description

A molecular pathologist thought he had finally identified a gene he had been working on. He submitted it to GenBank, the public database that contains every published DNA sequence. GenBank can identify similar genes and so is useful in trying to infer a new gene's function. But the database turned up more than 100 matches, a sign that something had gone terribly wrong. Indeed, each of those matches had in common a sequence that had been introduced by the commercial kit he had used to clone his gene.

The pathologist says that he found the error "entirely by accident," and that "there's a huge number of public sequences that are incorrect."

Source: Pennisi E. Keeping genome databases clean and up to date. *Science.* 1999;286:447–450.

Questions for Discussion

1. We know well that databases are dependent on those who build and maintain them, and that database design, construction, and maintenance raise

ethical issues. What special issues are raised when databases store biological or health information?

2. Who should be responsible for errors in very large and/or complex databases? What is to be done when errors are perpetuated, as perhaps by
commerical cloning kits? An error might be caught or missed, have no
effect or have a tragic effect—independent of the action that introduced
the error. Does the consequence of the error have moral significance?

3. Is database size or complexity an adequate excuse for errors?

*Note: Parts of this chapter are adapted from Goodman KW. Bioinformatics:
Challenges revisited.* MD Computing. *1999;16:17–20, with permission.*

Further Readings

Adams MD, Ventner JC. Should non-peer-reviewed raw DNA sequence data release be
forced on the scientific community? *Science.* 1996;274:534–536.

Annas GJ, Glantz LH, Roche PA. Drafting the Genetic Privacy Act: Science, policy, and
practical considerations. *J Law Med Ethics.* 1995;23:360–366.

Annas GJ. Privacy rules for DNA databanks: protecting coded 'future diaries'. *JAMA.*
1993;270:2346–2350.

Arena JF, Lubs HA. Computerized approach to X-linked mental retardation syndromes. *Am
J Med Genetics.* 1991;38:90–99.

Bentley DR. Genomic sequence information should be released immediately and freely in
the public domain. *Science.* 1996;274:533–534.

Boguski MS. Hunting for genes in computer data bases. *N Eng J Med.* 1995;333:645–647.

Bork P, Koonin EV. Predicting functions from protein sequences—where are the bottle-
necks? *Nat Genetics.* 1998;18:313–318

Dickson D. Open access to sequence data "will boost hunt for breast cancer gene." *Nature.*
1995;378:425.

Goodman KW. Ethics, genomics and information retrieval. *Comput Biol Med.* 1996; 26:223–
229.

Gostin LO. Genetic privacy. *J Law Med Ethics.* 1995;23:320–330.

Hilgartner S. Biomolecular databases. *Science Commun.* 1995;17:240–263.

Miller RA. Why the standard view is standard: people, not machines, understand patients'
problems. *J Med Philos.* 1990;15:581–591.

National Center for Biotechnology Information. GenBank, Availabe at: http://
www.ncbi.nlm.nih.gov/genome/seq/. Accessed October 9, 2001.

Waldrop MM. On-line archives let biologists interrogate the genome. *Science.*
1995;269:1356–1358

Wolf CR, Smith G, Smith RL. Pharmacogenetics. *BMJ.* 2000;320:987–990.

6

Evaluation: An Imperative to Do No Harm

" I GIVE UP. WHERE'S THE PATIENT ?"

Reprinted with permission from Sidney Harris.

In the future, health care providers will need to interact directly with clinical information systems to take full advantage of the capacity of these systems for organization of information and clinical practice suport. This will require physicians and other providers to change the way they have traditionally recorded, retrieved, and utilized clinical data. This chapter discusses the major advantages resulting from physician use of clinical information systems, the barriers to the acceptance and use of these systems, methods that may increase their acceptance and use, and the ethical imperative for evaluation.

Demands for information have intensified because of changes in reimbursement, the increase in prepaid contracts, and a focus on cost-effectiveness, continuous quality improvement, bench-marking, performance measurement, and clinical outcomes [1]. The current goals of cost containment and outcomes measurement cannot be met by older administrative systems that ignore the fact that providers determine as much as 75% of health care costs, and that are incapable of providing data on costs, quality, and patient outcomes [2]. In recognition of the critical importance that medical records play in the delivery of health care, the Institute of Medicine (IOM) has called for the development and implementation of computer-based patient records (CPRs) [3].

In response to the growing demand for health care data, business and industry sold $15 billion worth of information technology to health care organizations in 1997 [4]. While many of these products primarily support business decisions, the Joint Commission on Accreditation of Healthcare Organizations (JCAHO) lists 344 clinical-decision support systems on their approved list. These systems capture data and generate clinical performance indicators. To date, 100 of these systems, though commercially available, have not been tested or implemented by a provider of health care services [5]. The benefits of information systems that integrate all areas and functions of a health care delivery system into a network with a common computerized patient record have been demonstrated. However, a survey of 360 acute-care hospitals found that only 9% of the hospitals had computerized all areas and functions [6]. Not a single hospital had integrated its separate computer systems into a network. In ambulatory care settings, recent estimates indicate that computer-based records are in place in no more than 5% of group practices [7].

The slow acceptance and diffusion of clinical information systems is in sharp contrast to the rapid diffusion of most other health care technologies. Their lack of acceptance may reflect the fact that clinical information systems affect practice patterns and professional relations among groups within the organization [8]. The success of a clinical information system depends upon its integration into a complex organizational environment and its effective use. When the implementation of an information system interferes with traditional practice routines and fails to provide direct benefits to users, the system is not likely to be accepted by physicians and other health care providers [9]. Three decades of experience suggest that many of the systems that involve direct order entry by physicians will fail [10].

Benefits of Physician Use of Clinical Information Systems

Clinical information systems provide major benefits in direct support of patient care. These benefits include increased efficiency in managing clinical information and improved quality of care and cost savings through decision support and management of patient care.

Use of clinical information systems by physicians significantly improves the management of clinical information. The traditional paper chart readily provides only about a third of the data that the physician needs while providing patient care, and the chart's lack of structure makes it difficult to find specific information in a timely fashion. Deficiencies in accuracy and completeness of the medical record are also problems. Clinical information systems that create an electronic patient record facilitate the reporting, organizing, and locating of patient data [11].

A second function that clinical information systems perform is to support clinical decision-making. Decision support can take many forms: information about costs of clinical diagnostic procedures, guidelines, warnings or alerts about drug interactions and other clinical events requiring attention, or reminders to follow up on routine procedures, to name a few [12]. The influence of decision-support systems on physicians' clinical decisions is important because these decisions generate more than three quarters of all health care costs. There is evidence that alerts and reminders can significantly improve patient care [13], and that when physicians have been provided with decision support tools, there is improvement in resource utilization [14] and quality of care [15].

Clinical information systems also support the management of care. Coordination of the complex tasks involved in providing patient care is one of the most important organizational processes in the delivery of health services. For example, a system at LDS Hospital in Salt Lake City, Utah, creates an integrated CPR containing patient information from laboratories, pharmacy, nursing, surgery, intensive care units, and medical records. Much of this data is captured by point-of-care, real-time data entry through bedside terminals and workstations. This integrated clinical database is used along with expert system technology to support clinical decisions, to manage patient care, and to support CQI efforts [11].

Another potential benefit of clinical information systems is the detection and prevention of medical errors and adverse events. Between 44,000 and 98,000 Americans die each year because of medical errors, according to the National Academy of Science's Institute of Medicine. Medical errors may rank as the eighth leading cause of death [16]. Moreover, the total national cost of preventable adverse events is estimated to be as high as $37.6 billion per year, approximately 4% of national health expenditures in 1996 [17].

Properly designed clinical information systems have the potential to detect and prevent medical errors. A review of 25 clinical trials of clinical decision support systems concluded that these systems could improve physician performance and clinical outcomes [18]. Other studies suggest that computer generated alerts [19] and direct order entry by physicians can decrease medication errors and prevent injury to patients from adverse drug events [20, 21]. At the same time, clinical information systems that are not adequately designed, implemented, and evaluated have the potential to cause harm to patients. Therefore, there is an ethical imperative to thoroughly evaluate clinical information systems.

Barriers to Direct Physician Use
of Clinical Information Systems

Direct physician use of clinical information systems is essential to integrate clinical decision support and patient care management with older administrative functions such as charge capture and billing. Capture of clinical data and use of captured data for decision-making need to occur at the point of service. Despite evidence that clinical information systems can improve patient care, they have not been successfully implemented in a number of cases. Many of the medical information technologies available for decision support and care management have been added to legacy data systems designed to process claims and transactions. The level of clinical data provided to physicians and other care providers is not adequate for the treatment of patients [5]. Other barriers are primarily organizational [22].

Clinical information systems have not been successfully implemented because they have failed to demonstrate improvements in patient care [23] or operating cost savings [24]. In one instance, the infrequent exposure of residents to a newly installed clinical information system resulted in increased patient waiting time and staff work load [25]. Failure also occurred when a clinical information system was implemented in a group practice in which none of the physicians had intimate knowledge of the system or responsibility for decision making during implementation [26].

Furthermore, when clinical information systems interfere with traditional practice routines, they are not likely to be accepted by physicians. An information system installed at the University of Virginia Medical Center was implemented 3 years behind schedule at a cost 3 times the original estimate. The system was strongly opposed by physicians because it lacked medical staff sponsorship, altered traditional workflow patterns, changed the relations among professional groups, and adversely affected the medical education program [27,28].

The Impact of Implementing Information Systems

The introduction of computer information systems is a very different process from implementing new technology that gradually changes clinical processes and procedures. In general, the introduction of a comprehensive information system results in a discontinuous shift in technology that creates uncertainty and requires major adaptations in work processes [29]. Different occupational groups in health care organizations are affected differently by the introduction of new technology. Professional groups and organizational units strongly influence access to information, values, norms, and behavior of individual members [29–32].

Computerized information systems also affect job design, work force composition, and the coordination of workflow. The ultimate effects of the introduction of information technology depend upon-which groups select and control the system; the response of departments, professional groups, and individuals as they adapt to new work requirements and practice patterns; and the effect the system has on the ability of organizational groups and members to effectively perform their tasks. New information technology frequently results in the reassignment of tasks from one group to another. This process, by which the division of labor, social organization, and job designs change as a result of new information technology, is currently poorly understood [29].

Problems with New Technology

Frequently, there are unanticipated consequences of the introduction of new computer-based technology. Examples of unintended harm to patients abound. In one instance, a problem with the software that controlled the vertical position of a 3000-pound x-ray machine caused a patient to be crushed when the technician left the room [33]. A second example is the series of mishaps involving the Therac-25 in which errors in a radiation therapy machine resulted in the deaths of 3 patients and severe radiation burns to 3 others. The problem resulted from software errors, ambiguous error messages, a faulty microswitch, the absence of hardware interlocks, and inadequate testing and evaluation of the system [34].

Clinical information systems are far more complex and, therefore, have the potential for a wider range of unforeseen effects on providers and patients. For example, in October 1997, Oxford Health Plans, one of the largest managed-care plans in New York State, announced that problems with their newly installed computer system would result in a write-off of as much as $53 million for the third quarter [35]. Much of the write-off was used to cover bills owed to doctors and hospitals. Over a 9-month period, Oxford had implemented a major upgrade of its computer system but failed to adequately evaluate the capabilities of the new system. As a result, the implementation of system resulted in errors in payments to doctors and hospitals totaling hundreds of millions of dollars and major delays in billing subscribers for monthly premiums.

Kaiser Permanente, the largest managed care company in the United States, has begun a project that will move all its operations onto the Internet [36]. Kaiser plans to create computerized medical records for its 9 million members and to link all of Kaiser's 361 clinics and hospitals including 10,000 doctors, nurses and dentists. The project at one point was a year behind schedule and elicited resistance from some doctors who expressed concerns about the effect on physician–patient interactions of putting computers in the examining rooms. Kaiser's CEO admitted, "we may have underestimated the complexity of this undertaking."

More than 600 operational problems have been experienced with an EMR system put online at a regional medical center in April 1999. The staff has experienced problems retrieving medical records and in some cases the system failed to post laboratory and pathology reports to patient files. The system was installed without a gradual phase-in period during which the EMR could be evaluated [37].

Flaws in Information Systems

Information systems that are poorly designed and inadequately evaluated can cause harm to more than just stockholders and providers. At a hospital in a Western state, a 35-year-old woman was admitted to the emergency department for high fever and rigors. Laboratory tests were ordered, but when her blood pressure dropped, the woman was admitted to the intensive care unit (ICU) with a presumed diagnosis of sepsis. The ordering systems in the emergency department and the ICU were independent, and new orders had to be hand written by an ICU physician. The physician's order for an antibacterial medication (ampicillin) was misread by the pharmacy, which filled the prescription with an antiviral medication (acyclovir). The error was detected 2 hours after the medication was administered. The young woman, however, suffered irreversible brain damage as a result of the error [38].

Too narrow a focus on technical specifications can lead to the neglect of more systemic changes brought about by a new system and result in medical errors. In one case, a hospital hired consultants to design and set up a computer-based pharmacy system to help make dosages more accurate, decrease the time between prescription and treatment, and reduce costs [39]. The system was implemented after it passed all tests and specifications that were included in the original contract.

Within months, however, the hospital decided to return to its paper system because of potential risks to patients caused by medication errors. A problem was identified in the way the hospital staff interacted with the system, not in the software or hardware. The new system eliminated much of the oversight provided by staff using the old paper-based system. As a result, doctors, nurses, and pharmacy staff all blamed someone else for entering the wrong information when medication errors occurred. Because the computer system created a common database used by all hospital personnel, it was impossible to attribute errors to any one person or department.

Moreover, the system included data-handling and user-interface features from a warehouse inventory system. While the system met all of the original specifications, many of the problems involving errors, excessive time for data entry, and computer-generated advice that was unacceptable to physicians were not anticipated by the developers.

Ways to Anticipate and Prevent Problems

In order to anticipate and avoid errors with the potential for serious harm, a thorough evaluation of an information system is essential. In general, 10 questions need to be addressed [40]:

1. Does the system work as designed?
2. Is it used as anticipated?
3. Does it produce the desired results?
4. Does it work better than the procedures it replaced?
5. Is it cost-effective?
6. How well have individuals been trained to use it?
7. What are the anticipated long-term effects on how departments interact?
8. What are the long-term effects on the delivery of medical care?
9. Will the system have an impact on control in the organization?
10. To what extent do effects depend on practice settings?

Negative findings for any of these questions suggest a potential for medical errors.

A major impediment to the successful implementation of information systems in health care continues to be the lack of comprehensive evaluation [41]. A careful assessment can help organizations weigh the potential benefits of an information system for patient care against possible negative impacts for providers and patients. Moreover we have an ethical responsibility to conduct rigorous system evaluations, an ethical necessity implied by the clinician's duty to do no harm.

Overcoming Barriers to Physician Acceptance

Experience suggests several factors that may increase acceptance and use of clinical information systems by physicians [42]. First, broad physician involvement in the selection and implementation of the system from the outset is essential. Systems with no real sponsorship from the medical staff are likely to fail. In one effort to increase physician use of the hospital information system for direct order entry, the assistance of influential physicians was enlisted to encourage other members of the medical staff to use the system in practice [43]. Significant increases in the use of personal order sets to enter medical orders were observed on the experimental units. Not only did physicians on the experimental units increase the percentage of orders that they directly entered into the computer system, physician assistants and ward clerks or unit secretaries increased their use of computer-stored personal order sets to enter physician orders into the hospital information system. A major result of this increased use of the information system for direct order entry was a significant reduction in errors made during order entry.

High-level sponsors are needed when a new clinical information system is introduced into a practice setting. An example is the introduction of CompuHx, a computer-based health appraisal system, into the Department of Preventive Medicine at the Kaiser-Permanente Medical Care Program in San Diego, California [32]. CompuHx assists nurse practitioners and physician assistants who work under the supervision of the medical staff in gathering and recording patient information for diagnosis. The medical director was directly involved in the selection of the system and its initial implementation. An evaluation indicated that clinicians that were introduced to the system were willing to use it because they perceived that it improved their performance in providing better patient care. Nurse practitioners and physician assistants who used the system reported that they communicated more frequently with one another as well as with other staff who could assist them in performing their professional duties than did non-users. This frequent consultation and communication has potential benefits for patient care.

Second, it is essential to consider in advance how the system being introduced will affect routine practice patterns and professional relations. It is important to identify specific benefits that the information system provides to individuals and organizational groups. Physicians will use an information system if it helps them to provide better care for their patients. Benefits to the organization, in general, will not motivate physicians to change long-standing practice patterns.

There are several ways to assess the potential impact of a new clinical information system on practice patterns and thus avoid problems. One approach involved personal interviews with representatives of every hospital unit and a survey of every clinician [44]. Analysis of these data revealed a widespread lack of information about the benefits of the new system, so efforts were made to provide this information, training, and support services to physicians.

Third, health care organizations must be prepared to anticipate and manage a host of behavioral and organizational changes caused by the introduction of an integrated clinical information system. In one strategy to add direct order entry by physicians, a system was introduced into 1 new patient-care unit every 2 weeks [45]. Pharmacy technicians were available on a 24-hour basis for the first 3 or 4 days to assist physicians in learning to enter orders. By phasing in the system's implementation and anticipating problems, the hospital was able to reduce the number of negative experiences associated with the introduction of a new system and procedures.

The Future of Clinical Information Systems

In the future, integrated clinical information systems will be essential for the support of health care information management. Current trends point toward the accelerated growth of large health care systems. These organizations, in which most clinical decisions are made in clinics and physicians offices, will require accurate well-organized clinical data that can be made available rapidly at multiple sites. The successful implementation of information systems depends heavily upon integrating these systems into complex organizational settings. There is evidence that when physicians perceive that a computerized patient information system facilitates their practice, they will learn to use it, even if this requires changes in their practice behavior [46].

To ensure the use of clinical information systems, however, these systems must provide immediate benefits to clinicians. They must be flexible enough to allow the physicians to re-engineer their workflow in keeping with their personal practice styles. It is essential that physicians are involved in the entire selection and implementation process. This should include high level sponsorship by respected members of the medical staff, communication with individuals at all organizational levels to develop ways of incorporating system use into their practice, and establishment of lines of authority for making decisions and resolving conflicts. Finally, organizational changes that affect clinical practice should be made incrementally. Individuals need to be adequately trained to use the new system, to be given time to experiment with the system, and to be provided technical assistance.

Case Studies

Failure of Health Information Systems

Case 6.1: A Troubled Hospital Medical Information System

Description

A regional medical center invested $2.7 million in a hospital information system that stores patients' medical records. This EMR system is used by physicians to track the clinical progress of patients. The advantages of EMRs are that they allow faster access to a patient's clinical information by providers, they permit several providers to view a medical record simultaneously, and they can reduce medical errors, improve the quality of care, and reduce costs.

The system was installed in April 1999, without a gradual phase-in and evaluation period. Since implementation, there have been over 600 operational problems with the EMR, including difficulties in retrieving patients' medical histories and failure to post laboratory and pathology reports to patient files. Physicians and nurses express dissatisfaction with the system, complaining that it takes too much time to retrieve patient information that is frequently incomplete.

Physicians and nurses complained that it was faster to look up information in the patient's paper chart than to access the information through multiple computer screens, and that completing hospital rounds took much longer than before the system was installed. Nurses reported high levels of stress and low morale as a result of the problems experienced with the medical record system. In short, implementation of the system decreased efficiency rather than improving it.

When the hospital information system was installed, the hospital contracted to purchase at least 85% of its pharmaceutical, medical, and surgical supplies from the provider for the next 5 years. The $50 million contract included computer equipment and software as well.

Source: Henderson G. SRMC report: 600 problems with system since July. *Spartanburg Herald-Journal*, Spartanburg, South Carolina, March 22, 2000: A1.

Questions for Discussion

1. Was it appropriate for the hospital administration to introduce the new medical record system without an adequate evaluation and gradual phase in period?
2. What could the hospital administration have done to avoid the problems that the staff has experienced with the system?
3. Does the large multi-year contract with the provider of the system make it difficult to make major modifications to the system in response to staff complaints?
4. What should be done to address the stress and morale problems experienced by the nursing staff?

Case 6.2: A Borrowed Pharmacy System

Description

A small hospital was experiencing rapid growth. The vice president for records and automation and the chief pharmacist felt that the efficiency of the hospital pharmacy needed to be significantly improved. The hospital hired consultants to design and to implement a computer-based pharmacy system designed to make dosages more accurate, decrease the time between prescription and treatment, and reduce costs.

The system was implemented after it passed all tests and specifications that were included in the original contract. Since the chief pharmacist was directly involved in the project, he felt certain the new system would be beneficial to the hospital and staff.

Within months, however, the hospital decided to return to its paper system because of potential risks to patients caused by medication errors. A problem was identified in the way the hospital staff interacted with the system, not in the software or hardware. The new system eliminated much of the oversight provided by staff in using the old paper-based system. As a result, doctors, nurses, and pharmacy staff all blamed someone else for entering the wrong information when medication errors occurred. Because the computer system created a common database used by all hospital personnel, it was impossible to attribute errors to any one person or department. Moreover, the system included data-handling and user interface features from a warehouse inventory system. While the system met all of the original specifications, many of the problems involving errors, excessive time for data entry, and computer-generated advice that was unacceptable to physicians were not anticipated by the developers.

Source: Collins WR, Miller KW, Spielman BJ, Wherry P. How good is good enough? analysis of software construction and use. *Commun ACM.* 1994; 27:83–84.

Questions for Discussion

1. Did the hospital's technical staff, who were responsible for testing the new system against specifications, do enough testing? Could many of these problems have been foreseen and avoided by testing the new system?
2. How can the hospital regain the confidence and support of its professional staff and employees when they implement a new system?
3. What are the roles and obligations of the following groups in selecting, testing and implementing health information systems: the system provider or vendor, the organization purchasing or leasing the system, the care providers who will use the system, other individuals including patients who may be affected by the new system?

Case 6.3: Problems with a Billing and Payment System

Description

Oxford Health Plans, one of the largest HMOs in New York State, was founded in 1984. The company recruited a large network of providers, and members were offered the freedom to see any physician if they were willing to share the cost of their care. Unlike other managed care organizations, Oxford allowed patients with a specific medical problem, such as prostate cancer, to chose among a number of teams of specialists. The specialist had the freedom to provide patient care without securing approval from the insurance company. Alternative practitioners and procedures, such as acupuncture and herbal remedies, were also covered. Over 2 million New York residents joined the plan, resulting in annual revenue of $4 billion.

In October 1997, Oxford announced that problems with a newly implemented computer system would result in a write-off of as much as $53 million for the third quarter. Much of the write-off was used to cover bills owed to doctors and hospitals. Over a 9-month period, Oxford had implemented a major upgrade of its computer system but failed to adequately evaluate the capabilities of the new system. As a result, the system, once implemented, resulted in errors in payments to doctors and hospitals totaling hundreds of millions of dollars as well as major delays in billing subscribers for monthly premiums.

The problems have adversely affected stockholders, patients, and providers. In February 1998, the company's stock price dropped from over $90 per share to under $18. Patients have been affected by inconsistent decisions regarding coverage. In one case, a pregnant woman was certified as covered for delivery, but after she gave birth, the bill for $6500 was denied. In a second case, surgery for skin cancer was denied and classified as cosmetic surgery.

Providers responded angrily to the plan's failure to pay their bills on time and to disputes over the amount owed to providers. New York medical societies also complained, and some physicians hired attorneys on a contingency fee basis to collect outstanding payments. Physicians, on occasion, have even taken out their frustration with Oxford on patients. Reportedly one pediatrician told his/her office staff to give the cheaper lollipops to patients covered by the Oxford plan.

Source: Freudenheim MA. Troubled Oxford Health takes action. *The New York Times.*
    November 7, 1998:D1.

Questions for Discussion

1. What could Oxford Health Plans have done to avoid the problems that it experienced with its information system?
2. How can Oxford balance the demands of stockholders and providers in meeting outstanding financial obligations?
3. Would it be appropriate for Oxford to pass on the costs incurred in correcting the problems with the information system to subscribers in the form of higher premiums?

Case 6.4: Physician Resistance to an Internet-Based System

Description

Kaiser Permanente is the largest managed care provider in the United States. It consists of more than 300 hospitals and clinics and more than 100,000 physicians who provide care for 9 million subscribers. Kaiser had maintained paper medical records that were shuttled from hospital to clinics and back by couriers. Two thirds of the time, when patients were referred to a specialist, the patient's medical record was not available to the provider when the patient was seen. The costs and difficulty of providing care without a computer-based information system resulted in a loss of almost $300 million in 1998. Subscribers began to complain about cost cutting measures that reduce the quality of care they receive.

In 1998, to manage better the provision of care and reduce costs, Kaiser began to create an EMR for each of its members. The system uses the Internet to link clinics and hospitals. Providers can access the patient's medical record with a keyboard and a flat panel screen on the wall of the examining room. Each region has a central data base that contains the patient's clinical information including previous treatments, x-rays, and physician notes. When drugs are ordered, the system provides a list of lower cost alternatives. Another feature of the system allows patients to make appointments through the Internet. Kaiser has estimated that this feature could save $400,000 per year in the Northwest region alone.

There is some evidence that the system is functioning as anticipated. In the Northwest region, annual cost increases declined from 6% to 2.1% in 1999, and drug costs have been reduced by 20%. The system has reduced errors and duplicate tests. Kaiser estimates that, before the system was installed, as much as 15% of patient tests had to be repeated because of missing results or illegible provider notes.

However, the project had fallen behind schedule. A major problem has been physician resistance to using the Internet-based system. Some physicians argued that having the computer system in the examining room interferes with physician-patient interaction. They report feeling awkward talking to patients and, at the same time, using the system to access information or enter notes. Other physicians have complained about system features that provide advice or suggest alternative medications.

Source: Gantenbein D, Stepanek M. Kaiser takes the cyber cure. *Business Week Online.* February 7, 2000. Available at http://www.businessweek.com/2000/00_06/b3667061.htm. Accessed October 9, 2001.

Questions for Discussion

1. Are there ways to ensure that the introduction of terminals in the examining room does not interfere with the physician-patient interaction?
2. How can Kaiser overcome physician opposition to the advice that the information system provides when the physician writes orders?

Case 6.5: Unanticipated Consequences of a Physician Order Entry System

Description

A university medical center implemented new financial and accounting systems and a medical information system (MIS). After the accounting programs were installed, the MIS was implemented in 2 phases; first, administrative functions, then clinical functions. The system was designed to include all physician orders, laboratory results, and radiology findings.

A great deal of organizational stress was engendered by the implementation of the system. The MIS changed traditional work patterns and relations among professional groups, and reactions of the different professional groups varied.

Since physicians were supposed to enter their own medical orders directly into the MIS through computer terminals placed throughout the hospital, they felt that functions previously performed by nurses had been transferred to them. Nurses gained some autonomy from the physicians but felt left out of the ordering process. House staff had negative feelings toward the new system because it required that all verbal orders had to be signed by the attending physician before the orders could be implemented. Pharmacists felt that the system had a positive effect in eliminating illegible and incomplete medication orders.

Resistance to the system developed among the medical staff because the changes in their practice behavior resulting from the implementation of the system were attributed to an administrative decision imposed from outside the medical staff. Physicians perceived the system as providing improved financial and managerial information to the medical center administration but providing few benefits to attending physicians.

The introduction of the MIS also had unanticipated effects on medical education as well. Residents complained that use of the MIS took too much time, and medical students could no longer activate pharmacy and laboratory orders. Not only was the assistance that medical students provide to residents compromised, but residents had less time to teach medical students. Attending physicians generally did not enter their own orders into the MIS, resulting in a situation in which residents knew more about using the system than their mentors did.

Sources: Massaro TA. Introducing physician order entry at a major academic medical center: I. Impact on organizational culture and behavior. *Acad Med.* 1993;68:20–26; and Massaro TA. Introducing physician order entry at a major academic medical center: II. Impact on organizational culture and behavior. *Acad Med.* 1993;68:25–30.

Questions for Discussion

1. How might administrators have reduced organizational stress when the new system was introduced?
2. What are the major concerns and expectations of each of the following groups when a medical information system is introduced: administrators, physicians, residents, nurses, medical students, and pharmacists?
3. Could some of the conflict that arises between different professional groups be worked out through a coordinating committee on which all user groups are represented? Who should be represented on this committee?

Case 6.6: Automation of Inefficient Processes

Description

Many electronic medical record systems focus on documentation that will guarantee payment rather than lead to improved clinical decision making. Systems developed for use with direct patient care frequently automate existing processes such as scheduling patient appointments and capturing charges for billing.

In one instance a large medical insurance company automated its claims processing system. Consultants and information technology professionals studied and then automated what the claims processors did. Little attempt was made to improve the process, take full advantage of automation or to anticipate future system requirements. The budget for developing the medical knowledge support system was $30 million.

The system included decision support tools and data on standards of care. The purpose of the system was to reduce variation in practice and to document clinical decisions to advance the development of standards of care. Medical information was to be provided to physicians in real time while they were providing care for a patient. Additionally, physicians were to retain control of the database incorporated into the information system. After spending $80 million on the claims processing component of the system, management abandoned the project. Development of the medical knowledge support component was never begun since the computer-based system locked in the original inefficient process. Eventually the information system could not meet the data requirements of managed care.

Source: RH Strube, personal communication.

Questions for Discussion

1. How can organizations avoid the problem of automating existing inefficient processes when they develop and/or implement an information system?
2. Was the development of the medical knowledge support too ambitious a project for the organization to undertake at the same time it was automating the claims process? Could problems have been avoided by implementing one project first, followed by the second project?
3. Can physicians retain control over the knowledge base of decision support systems in a managed-care environment?

Case 6.7 Failure of a "Best of Breed" Information System

Description

An Australian health department provides public hospital and community services for a population of more than 7 million people. The health system is composed of 17 autonomous geographically based units called Area Health Services. The units are under the direction of a central department that sets policies and monitors performance. In 1989, a decision was made to implement a suite of state-of-the-art information systems. In order to maximize the chance of successful implementation, the best available systems for finance, pathology, and clinical services were selected from around the world. The clinical system had been implemented in about 100 other hospitals worldwide. Even though the system's functionality had been well established, implementation of the clinical system was not successful, and it was withdrawn from all 5 pilot sites.

An analysis of the failure to implement the system pointed to a number of organizational factors. First, the health system can be characterized as a "professional bureaucracy"—that is, the professionals who perform the critical functions that drive the organization are autonomous, and the administrative component of the organization is managed as a bureaucracy. Managers, who were in charge of the implementation of the new information system, had little control over the doctors, and the doctors saw little value in the new system.

Second, the administration assumed that because the system worked in other settings it could be made to work in their health system. This turned out not to be the case because of differences in work practices. In other settings in which the system was implemented, clerks entered clinical information, but in this setting, doctors were expected to enter medical orders directly into the system. The system was not particularly user friendly, and many doctors were not adept at using it. Though the system was promoted with a list of general benefits such as support of patient services, many of the benefits were too vague for the staff to relate to and not credible enough to inspire the commitment of the clinicians.

Third, decision-making was spread among many people with different interests and expertise. Funding came from the government, but implementation was undertaken at each site. Thus, it was difficult to resolve problems as they arose.
Source: Southon G, Sauer C, Dampney K. Lessons from a failed information system initiative: issues for complex organizations. *Int J Med Inf.* 1999;55:33–46.

Questions for Discussion

1. When implementing an information system simultaneously in multiple institutions, how can a balance be maintained between local control and needs and the need for commonality?
2. Would the health system have been more successful if it had adopted a strategy of incrementally enhancing the old system instead of introducing a totally new information system all at once? What features of the system should be enhanced first?
3. How could the management team responsible for implementing the information system have gone about securing commitment to the system on the part of the physicians?

Case 6.8 No OSCAR

Description

A Canadian hospital installed the Technicon Data System (TDS) online systems for communications and records (OSCAR). The hospital's management decided that having physicians input their own medical orders could significantly reduce data-entry costs. Resistance to the new system grew rapidly, because under the new policy, residents had to make more than 70% of all medical order entries. Residents said that OSCAR created stress because it changed the way that work was done in the teaching hospital, requiring them to enter orders twice, once at the bedside, and once at a terminal, and adding an additional 30 minutes to the time that it took to make rounds. There was conflict between residents and nurses over who was to enter medical orders. The implementation of OSCAR also raised issues of patient confidentiality, since all patients' medical charts could be accessed by anyone on the hospital staff with an access code. In the end, the residents signed a petition requesting that direct order entry be made voluntary.

The new system also resulted in serious errors. In one instance, OSCAR reordered blood transfusions and laboratory work on patients in the hospital. The new orders were copies of previous orders that the computer system printed and signed with physicians' personal electronic signature. A resident noticed the error, and the hospital advised nursing stations to disregard the OSCAR ordering system. In a second instance, a patient received the wrong antibiotics. A doctor had called up his patient's chart on OSCAR and entered an order for an antibiotic, but OSCAR had pulled up the wrong patient's medical record becaue the patients' surnames were the same, but their first initials were different.

Source: Williams LS. Microchips versus stethoscopes: Calgary hospital, MDs face off over controversial computer system. *Can Med Assoc J.* 1992;147:1534–1547.

Questions for Discussion

1. Hospital administrators are generally removed from medical practice and education and may not appreciate the stress that staff experience when an information system like OSCAR is implemented. How can the administration anticipate some of the major stresses that staff may experience during implementation?
2. What could the hospital's administration have done to better prepare the house staff to use the OSCAR system?
3. How might the negative effects of the introduction of OSCAR have been minimized?

Case 6.9: An Unsuccessful Computerized Medical Record System

Description

The Family Practice Center at an academic medical center is a teaching clinic that sees approximately 20,000 patients each year. Prior to implementation of a computerized information system, patient records were kept in manila folders. The center decided to computerize the medical record in order to achieve a number of reported advantages including better legibility; automatic generation of problem lists, medication lists, and physician and patient reminders; access to medical records from remote sites; development of a clinical data base for research purposes; and better monitoring of the training experiences of residents and medical students.

The system chosen was the Computer Stored Ambulatory Record (COSTAR) system. COSTAR is the system most widely used in ambulatory settings in the

United States. Its functions include medical record keeping, billing appointment scheduling, generation of insurance forms, and revenue analysis. Physicians recorded information on data-entry forms that required that almost all information be recorded in coded format. Data-entry clerks typed information recorded on the data-entry forms into the computer system, and entered results of x-rays, consultation reports, electrocardiograms, and other tests. Physicians and other authorized staff accessed the system using personalized identification codes.

Because the administration of the medical center required that the cost of the system not exceed the cost of the paper-based patient record system it replaced, the system was terminated after 4 months when patient-care revenue could not cover the cost of operating the new information system. A number of problems were identified. Data-entry clerks were unable to enter data from the physicians' data-entry forms within 24 hours, and by the end of the 4 months, the entry of telephone messages, consultation reports, and diagnostic test results was as much as 3 months behind schedule. Moreover, during peak periods, the system's response was so slow that physicians would abandon attempts to retrieve patient data. An audit of the medical records revealed that approximately 15% of the records contained errors attributable to difficulties in reading the physicians' notes on the data-entry forms. The Medicare administration refused to accept computer-generated medical records as evidence of direct contact between patients and physicians without handwritten documents signed by the physician.

Source: Dambro MR, Weiss BD, MCClure CL, Vuturo AF. An unsuccessful experience with computerized medical records in an academic medical center. *J Med Educ.* 1988;63:617–623.

Questions for Discussion

1. Is four months an adequate period of time to assess the cost-effectiveness of a new information system? If not, how much time is needed? Why?
2. In what ways were the administration's expectations regarding the cost and the advantages of the computerized medical record system unrealistic?
3. Would new technologies that facilitate data entry by reducing transcription errors and the number of personnel needed to enter data into the system make the system more cost-effective?

Case 6.10: An Abandoned Computerized Medical Record System

Description

A pediatric primary care group practice at a medical school also implemented the COSTAR system. The practice consisted of 4 general pediatricians, 21 residents in pediatrics, 5 nurses, 5 clerical workers, and 1 secretary. Approximately 80 third-year and fourth-year medical students rotated through the practice over the course of 18 months.

Medical data were captured on 2 encounter forms, one for telephone interactions and one for office visits. Information was coded and entered by clerks in a specific sequence: present illness, past medical history, results of the physical examination, diagnoses, problems, tests ordered, medications, and follow-up plans and schedule. A record that included a summary of the patient's medical information from the 3 most recent encounters was printed at each subsequent office visit. The hospital's laboratory system was also interfaced with the system so that test orders and results were immediately transferred. Authorized providers could gain access to patient information from terminals throughout the clinic and hospital as well as from homes with personal computers that had dial-in capability. Appointments with attending physicians were scheduled through the computer system, and the ambulatory-care quality-assurance program also used the information contained in the system.

After 18 months, the pediatricians, nurses and clerical staff expressed satisfaction with the computer-based record system, preferring it to the paper-based medical record that it replaced. Users felt that the system resulted in improved availability of information, office efficiency, and patient care. Residents were also positive about the system, once they had adapted to it. The EMR enhanced the clinic's ability to document residents' training experiences and monitor quality of care. However, the medical center administration decided to discontinue the system because of the cost, a lack of consensus about the value of the system among physicians in different clinics, and slow implementation of the new medical record system in other clinics of the medical center.

Source: Chessare JB, Torok KE. Implementation of COSTAR in an academic group practice of general pediatrics. *MD Computing.* 1993;10:23–27.

Questions for Discussion

1. When a computerized medical record system like COSTAR is being simultaneously implemented at multiple locations, it is unrealistic to expect implementation to proceed at the same rate at all locations. How can an institution coordinate the implementation at multiple sites?
2. One of the problems experienced in other hospital clinics was that physicians who had not been involved with the planning of the project did not appreciate its potential benefits. How could this problem have been avoided?
3. Another problem was that the management structure of the project was diffuse. No one physician had the authority to make day-to-day decisions needed for implementation. As a result, some physicians refused to participate in the implementation or to use the COSTAR system. Could this problem have been foreseen and avoided? How?

Case 6.11: The Effects of Computerizing
Medical Records on Clinic Function

Description

The COSTAR system was installed in the teaching clinic of the Internal Medicine Department at a university medical school. Resident physicians who staff the clinic were assigned to a control group, which was allowed access to the paper-medical record only, or to a study group, which was allowed access to both the paper chart and the computerized patient record. After a patient was seen in the clinic, clerks entered provider encounter form data into the patient's electronic record, and a duplicate was filed in the patient's paper chart.

At the end of the 9-month study, a questionnaire was used to measure staff opinions regarding the effect of the system on clinic function and user attitudes toward the COSTAR medical record system. Thirteen aspects of the effect of the system were measured, including accuracy and organization of information; the time required to find patient information; the ability to find current medications, laboratory and x-ray results; and the general availability of medical records. The effect of the system on clinic efficiency was measured as the rate of patient flow through the clinic and the influence of the system on telephone management of patients.

The nurses and clerical workers strongly preferred all aspects of the EMR to the paper chart. Their positive response to the system was based on substantial improvement in the availability of medical records and time saved in telephone management of patients. Using the conventional paper chart, nursing and clinic personnel in the control group waited for information for a total of 5590 hours before making decisions related to telephone calls. In 2880 telephone contacts, they could not act with complete information. In contrast, computerized medical records were instantly available to staff in the study group over 99% of the time. Resident physicians, however, were less enthusiastic about the system. Their response was mitigated by the need to learn to use a new computer system while taking care of patients in a busy clinic. Although the analysis indicated improvements in performance among physicians in the study group, patients seen by these residents spent more time in the clinic's waiting room and more time in the clinic as a whole than the control group.

Source: Campbell JR, Givner N, Seelig CB, Greer AL, Patil K, Wigton RS, et al. Computerized medical records and clinic function. *MD Computing*. 1989;6:282–287.

Questions for Discussion

1. Could the increased time that patients spent in the clinic be attributed to the difficulties that residents experienced in learning to use the new information system? Would the time patients spend in the clinic decrease as the residents learned to use the system?
2. Would it be possible to reduce the system's adverse effects on patient flow by changing the scheduling system in the clinic? How?

3. When implementing an EMR system, differential benefits result for different occupational groups (i.e., physicians, nurses, clerical staff). What are the major benefits for each of these groups?

Case 6.12: Electronic Prescriptions

Description

A Veterans Affairs medical center became aware a of a serious budget shortfall, amounting to almost 10% of the facility's budget. Almost half of the shortfall was due to increased prescription drug costs. Mainly, the problem appeared to be an indirect result of implementation of a new local version of an EMR system called the Electronic Chart. The medical center had hired several highly talented programmers to create the Electronic Chart, which enabled physicians to store progress notes electronically; to order tests, procedures and consults; and to access clinical and demographic information from a personal computer. After an early version of a program that permitted electronic entry of prescriptions was resisted by physicians, the pharmacy began to print out an "Action Profile" listing all the patient's medications, dosages, and remaining refills, and providing space to list new prescriptions. Physicians readily accepted this procedure, because all they needed to do was to sign their name below the preprinted information rather than rewrite all prescriptions in full each time. An unintended consequence of the new system was that busy physicians began to sign all prescriptions listed on the Action Profile resulting in increased prescription costs.

The administration and the medication use committee wanted to enhance the local electronic prescription order entry system in order to control costs of prescription drugs. However, the Veterans Health Administration (VHA), which comprises a national network of medical centers, outpatient clinics, domiciliaries, and community counseling centers, has mandated that all VA facilities implement a national electronic prescription order system. It is estimated that implementation will take 2 years.

Use of the mandated electronic prescription order system will necessitate abandonment of the popular Action Profile. Furthermore, the mandated system does not provide information about which drugs are available on the formulary, does not provide dosing information and does not provide physicians with a global view of the patient's medications. The prescription is transferred to the pharmacy by e-mail, where it is printed out and must be re-entered by a pharmacy technician. The programmers, who are being asked to make further modifications to the local system, believe that if they make the requested modifications, it will be even more difficult for the local system to interface with the national system. Support for the project among the medical staff varies. Moreover, house staff who rotate through the center comprise 40% of the medical staff and generate a significant proportion of the prescriptions, presenting a special problem because of the short time that they are on the services.

Source: Carpenter J, Hiruki T, Krall M, Smith D. Fix pharmacy! Case presentation at: Oregon Health Sciences University; March 11, 1999; Portland, Ore.

Questions for Discussion

1. It appears that the implementation of an enhanced local electronic prescription order system may exacerbate the stresses among the staff of the center. How can this be reduced or avoided?
2. Will the implementation of an enhanced local electronic prescription order system ultimately make it more difficult to implement the national system?
3. How can the center insure physician support if it goes ahead with the decision to enhance the local electronic prescription order system?

Case 6.13: Programming Error Results in Overpayment of Doctors

Description

Aetna, a health insurance company, offers the National Advantage Plan (NAP). Under this plan, Aetna negotiates discounted rates from health care providers and passes on the benefits to employers and their employees. The plan is a traditional medical indemnity program covering 80% of covered medical expenses after a deductible.

An error in the software of Aetna's information system resulted in overpayment of doctors for services that they were suppose to provide at a discounted rate. Providers billed patients at a higher rate than the discounted rate.

Source: Martinez B, Hensley S. Aetna acknowledges programming error. *The Wall Street Journal.* August 25, 2000:B2.

Questions for Discussion

1. Is there any way that Aetna could have prevented this error from occurring? How?
2. Are physicians who overbilled patients as a result of the error ethically obligated to refund the overpayment to patients? If physicians do not refund the overpayment by patients, is Aetna ethically obligated to do so?
3. Is this error likely to cause patients to mistrust their health care providers? How can Aetna regain their trust?

Information Systems That Harm Patients

Case 6.14: Malfunction of the Therac-25

Description

Therac-25 was a medical linear accelerator built by Atomic Energy of Canada

Limited (AECL) and designed to destroy cancer cells by irradiating them with either electrons or photons at high energy levels. The system was designed to be largely computer controlled. Consequently, safety was maintained in most instances by software, although the system had protective circuits to monitor the electron beam. Some of the software from earlier versions of the linear accelerator was reused. Eleven Therac-25s were installed in practice settings in the United States and Canada.

Over a 2-year period, 6 instances occurred in which patients received massive overdoses of radiation. Three of these patients died from severe radiation burns. After the first accident, engineers who were contacted at AECL maintained that improper scanning with the machine was impossible because of safety features built into the software. There was no subsequent investigation of the accident by the company, and it was not reported to the FDA until accidents occurred at other sites. Since health care institutions and professionals were not required to report incidents of equipment malfunction to manufacturers until the law was amended in 1990, other users of the equipment were unaware of problems with the Therac-25.

After a second accident resulted in the death of a patient, AECL tried unsuccessfully to reproduce the malfunction that had resulted in the fatal radiation burn. Nevertheless, in a report to the FDA, it claimed to have significantly improved the safety of the machine. In a letter, AECL advised 4 users in the United States to check the position and settings of certain features of the Therac-25 but did not mention that patients had been injured, one fatally. When a third serious radiation burn occurred (at still another site) hospital staff ascribed the problem to unknown causes, and AECL told the hospital that no machine malfunction had occurred. The hospital was not informed of incidents that had occurred at other sites using the Therac-25. After 6 accidents, 3 fatal, over a 2-year period, the FDA declared the Therac-25 defective and demanded that the manufacturer file a corrective action plan with the agency and notify all users of the Therac-25 of the problems. A subsequent investigation determined that the problem resulted from software errors, ambiguous error messages, a faulty microswitch, the absence of hardware interlocks, and inadequate testing and evaluation of the system.

Source: Leveson NG, Turner CS. An investigation of the Therac-25 accidents. *Computer.* 1993; 26:18–44.

Questions for Discussion

1. Many medical errors are often caused by system problems; that is, they result from complex interactions between many aspects of the system and its users. Is it possible to anticipate all problems in advance and to prevent errors from occurring?
2. Do users of information technology have an obligation to inform providers of the software, regulatory agencies, and other users of problems with the technology that may result in harm to patients?
3. Are there ways to prevent adversarial relationships developing among vendors of information technology, health care administrators who purchase or

lease the technology, clinicians that use it, patients and their attorneys, and government agencies? Can procedures or regulations be found that encourage identification of problems that may cause errors and early notification of all parties who are potentially affected by the problems?

4. Does the use of comprehensive EMR systems with decision support diminish accountability for resulting errors and adverse events? How can accountability be determined in the event of an error?

Case 6.15: Failure to Integrate Information Systems

Description

To compete for contracts with high-tech firms, a 450-bed hospital with a medical staff of 650 has purchased a number of private clinics and has instituted its own managed care insurance plan. It has a medical information system comparable to many other medical centers, but the system currently lacks physician order entry (POE) capability. Since at least one other hospital in the area has plans to implement a POE system, the CEO has informed corporations considering a contract with the medical center that a POE system will be implemented shortly.

An earlier attempt to implement a nursing documentation system against the wishes of the medical staff failed. More recently, another information system was implemented despite opposition from the medical staff. Physicians contend that managed care has increased their work load, decreased their incomes, and reduced their autonomy, and view computer-based information systems as adding to their workload.

The medical center is under pressure to implement a comprehensive medical information system because of a recent incident. A 35-year-old woman employed by a high-tech firm was admitted to the emergency department with a high fever. A physician ordered laboratory tests, and when her blood pressure dropped, the woman was admitted to the ICU with a presumed diagnosis of sepsis. The ordering systems in the emergency department and the ICU were independent, and though a physician ordered an antibacterial medication (ampicilllin), the pharmacy filled the prescription with an antiviral medication (acyclovir). The error was detected 2 hours after the medication was administered, but the young woman suffered irreversible brain damage.

As a result of the earlier failure of an information system and the lawsuit resulting from the incident in the emergency room, the Board of Directors for the hospital created the position of Medical Director of Information Services from among the medical staff to have responsibility for the new information system. The CEO is pressing for rapid implementation of the POE system but the Medical Director of Information Services fears that the medical staff is not ready for the implementation and that a hasty implementation without staff preparation and evaluation may result in medical errors.

Source: Ash JS, Anderson JG, Gorman PN, Zielstorff RD, Norcross N, Pettit J, et al. Managing change: Analysis of a hypothetical case. *J Am Med Inform Assoc.* 2000:7:125–134.

Questions for Discussion

1. Rapid implementation of a POE system, without adequate participation of the medical staff, is likely to result in failure of the system and potentially adverse events. How could the medical center go about enlisting physician support for and participation in the implementation of the physician order entry system?
2. Should the Medical Director of Information Systems refuse to implement the POE system without active support and participation by the medical staff? Could a revised timetable be negotiated with the CEO and Board of Directors of the Medical Center?
3. Does the medical center have a responsibility to consider the impact of the new system on the workload and practice of its medical and nursing staff before implementation?

Case 6.16: Concerns about the "Bedside Assistant"

Description

A firm that develops medical information systems contracted with a large teaching hospital to develop a decision support system. The system is termed a "bedside assistant." Physicians will be able to gain immediate access to information contained in the patient's chart through computer terminals located in each patient's room. The information contained in the system will alert physicians to potential side effects of medications and drug interactions. The decision support system will propose differential diagnoses and critique treatment plans.

The hospital administration wants the bedside assistant implemented as soon as possible, since a lot of money has been invested in the system development. Administrators anticipate that providing physicians with immediate access to the medical information and decision-support tools will lead to better diagnoses, improved treatment, and reduced costs. By recording all of a patient's medications, checking for side effects and drug interactions, and evaluating medications relative to the patient's condition, the hospital hopes to prevent potentially dangerous adverse drug events.

The engineers developing the decision support, however, have major concerns about its safety. They worry that the system may malfunction because of a bug in the software or because incorrect data has been entered into the patient's record. Though the development team consulted published sources and medical experts in the field there is no guarantee that they fully understand the information on which the system is based.

The user interface presents other potential problems. While it has been tested under controlled conditions, its performance under actual practice conditions is unknown. Data or commands may be incorrectly entered into the system, and data commands and advice provided by the system may be misinterpreted. It is clear to the developers that the system should be tested by running it in parallel with the existing system for some time before it is fully implemented. However, the hospital administration is not likely to agree to such a proposal because of the costs.

Source: McFarland MD. Ethics and the safety of computer systems. *Computer.* 1991;24:72–75.

Questions for Discussion

1. How long a testing period would be required to determine the safety and reliability of the bedside assistant system?
2. What level of reliability is adequate for a clinical decision support system?
3. Since no information system will be perfect, what levels of risk are acceptable?
4. What safeguards need to be built into the systems to protect patients from medical errors?
5. Since there are risks involved to patients, to what extent should physicians who will be using the new decision support system be involved in designing, testing, and implementing the new information technology?

Note: Portions of this chapter are taken from the following sources: Anderson JG. Clearing the way physician's use of clinical information systems. Commun ACM. 1997;40:83–90; Anderson JG. Increasing the acceptance of clinical information systems. MD Computing. 1999;16;62–65; and Anderson, J.G. Evaluating clinical information systems: a step toward reducing medical errors. MD Computing. 2000:17:21–23, with permission.

References

1. Kleinke JD. *Bleeding edge: the business of health care in the next century.* Gaithersburg, MD, Aspen Publishers; 1998.
2. Millenson, ML. *Demanding medical excellence: doctors and accountability in the information age.* Chicago: University of Chicago Press; 1997.
3. Dick, RS, Steen EB, Detmer DE, eds. *The computer-based patient record.* Rev. ed. Washington, DC: Institute of Medicine, National Academy Press; 1997.
4. Gallo AC, Lee VJ. *Health Care Information Technology: Keeping Health Care Wired,* Research report. Baltimore: Alex Brown; 1998, 1.
5. Kleinke JD. Release 0.0: Clinical information technology in the real world. *Health Affairs.* 1998;17:23–38.
6. Furfaros C, Muchoney K, Anania-Firouzan P. CPR by the year 2000: a myth? *Healthcare Inf.* 1996;13:45, 47.

7. McCormack J. When will smaller medical groups discover computers? *Health Data Manage.* 1997;5:50–52,54,56,58,60,63.
8. Anderson JG. Computer-based patient records and changing physician practice patterns. *Topics Health Inform Manage.* 1994;15:10–23.
9. Schoenbaum SC, Barnett GO. Automated ambulatory medical records systems: An orphan technology. *J Technol Assess Health Care.* 1992;8:598–609.
10. Sittig DF, Stead WW. Computer-based physician order entry: The state of the art. *J Am Med Inf Assoc.* 1994;1:108–123.
11. Kuperman GJ, Gardner RM, Pryor TA. *HELP: a dynamic hospital information system.* New York: Springer-Verlag, 1991.
12. Miller RA. Medical diagnostic decision support systems—past, present, and future: A threaded bibliography and brief commentary. *J Am Med Inform Assoc.* 1994;1:8–27.
13. Johnston ME, Langton KB, Haynes RB, Mathieu A. Effects of computer-based clinical decision support systems on clinician performance and patient outcome: a critical appraisal of research. *Ann Inter Med.* 1994;120:135–142.
14. Tierney VM, Miller ME, Overhage JM, McDonald CJ. Physician inpatient order writing on microcomputer workstations: Effects on resource utilization. *JAMA.* 1993;269:379–383.
15. Rind DM, Safran C, Phillips RS, et al. The effect of computer-based reminders on the management of hospitalized patients. In: Clayton P, ed. *Proceedings of the Fifteenth Annual Symposium on Computer Applications in Medical Care.* Los Angeles, IEEE Computer Society; 1991:28–32.
16. Kohn LT, Corrigan JM, Donaldson MS, eds. *To err is human: building a safer health system.* Washington, DC: National Academy Press, 1999.
17. Thomas EJ, Studdert DM, Newhouse JP, et al. Costs of medical injuries in Utah and Colorado. *Inquiry.* 1999;36:255–264.
18. Hunt DL, Haynes RB, Hanna SE, Smith K. Effects of computer-based clinical decision support systems on physician performance and patient outcomes: A systematic review. *JAMA.* 1998;280:1339–1346.
19. Raschke RA, Gollihare B, Wunderlich TA, et al. A computer alert system to prevent injury from adverse drug events. *JAMA.* 1998;280:1317–1320.
20. Anderson JG, Jay SJ, Anderson MM, Hunt TJ. Evaluating the potential effectiveness of using computerized information systems to prevent adverse drug events. In Masys D, ed. *The emergence of internetable health care—systems that really work: Proceedings of the 1997 AMIA annual symposium.* Philadelphia: Hanley S. Belfus, 1997:228–232.
21. Bates DW, Teich JM, Lee J, et al. The impact of computerized physician order entry on medication error prevention. *J Am Med Inf Assoc.* 1999;6:313–321.
22. Lorenzi NM, Gardner RM, Pryor TA, Stead WW. Medical informatics: The key to an organization's place in the new health care environment. *J Am Med Inf Assoc.* 1995;2:391–392.
23. Barnett GO, Winickoff RN, Morgan MM, Zielstorff RD. A computer-based monitoring system for follow-up of elevated blood pressure. *Med Care.* 1983;21:400–409.
24. Dambro MR, et al. An unsuccessful experience with computerized medical records in an academic medical center. *J Med Educ.* 1988;63:617–623.
25. Campbell JR, Givner N, Sellig CB, Greer AL, Patil K, Wigton RS, et al. Computerized medical records and clinic function. *MD Computing.* 1989;6: 282–287.

26. Chessare JB, Torok KE. Implementation of COSTAR in an academic group practice of general pediatrics. *MD Computing.* 1993;10:23–27.
27. Massaro TA. Introducing physician order entry at a major medical center: I. Impact of organizational culture and behavior. *Acad Med.* 1993;68:20–25.
28. Massaro TA. Introducing physician order entry at a major medical center: II. Impact on medical education. *Acad Med.* 1993;68:25–30.
29. Anderson JG, Aydin C. Theoretical perspectives and methodologies for the evaluation of health care information systems. In: Anderson JG, Aydin CE, Jay SJ, eds. *Evaluating health care information systems: methods and applications.* Thousand Oaks, Calif: Sage;1994:5–29.
30. Anderson JG, Jay SJ. Computers and clinical judgment: The role of physician networks. *Soc Sci Med.* 1985;20:969–979.
31. Anderson JG, Jay SJ, Schweer HM, Anderson MM, Kassing D. Physician communication networks and the adoption and utilization of computer applications in medicine. In: Anderson JG, Jay SJ, eds. *Use and impact of computers in clinical medicine.* New York: Springer-Verlag; 1987:185–199.
32. Aydin CE, Anderson JG, Rosen PN, Felitti VJ, Weng HC. Computers in the consulting room: a case study of clinician and patient perspectives. *Health Care Manage Sci.* 1998;1:61–74.
33. Friedman B, Kahn PH. Educating computer scientists: linking the social and the technical. *Commun ACM.* 1994;37:65–70.
34. Leveson N, Turner C. An investigation of the Therac-25 accidents. *Computer.* 1993;26:18–41.
35. Freudenheim M. A troubled Oxford Health takes action. *The New York Times.* November 7, 1997: D1.
36. Gantenbein D, Stepanek M. Kaiser takes the Cyber cure. *Business Week.* February 7, 2000. Available at: http://www.businessweek.com/2000/00_06/b3667061.htm. Accessed October 9, 2001.
37. Henderson G. SRMC report: 600 problems with system since July. *Spartanburg Herald-Journal,* Spartanburg, South Carolina, March 22, 2000: A1.
38. Ash JS, Anderson JG, Gorman PN, Zielstorff RD, Norcross N. Pettit J, Yao P. Managing change: analysis of a hypothetical case. *J Am Med Inf Assoc.* 2000;7:125–134.
39. Collins WR, Miller KW, Spielman BJ, Wherry P. How good is good enough? An ethical analysis of software construction and use. *Commun ACM.* 1994;27:81–91.
40. Anderson JG, Aydin CE. Technology assessment of health care information systems. *Int J Technol Assess Health Care.* 1997;13:380–393.
41. Anderson JG, Aydin CE. Evaluating medical information systems: Social contexts and ethical challenges. In: Goodman KW, ed. *Ethics, computing and medicine: informatics and the transformation of healthcare.* New York: Cambridge University Press; 1998: 57–74.
42. Anderson JG. Increasing the acceptance of clinical information systems. *MD Computing.* 1999;16;62–65.
43. Anderson JG, Jay SJ, Perry J, Anderson MM. Modifying physician use of a hospital information system: A quasi-experimental study. In: Anderson JG, Aydin CE, Jay SJ, eds. *Evaluating health care information systems: methods and applications.* Thousand Oaks, Calif: Sage 1994:276–287.
44. Martin-Baranera M, Planas I, Palau J, et al. Assessing physicians' expectations and

attitudes towards hospital information systems. The IMASIS experience. *MD Computing*. 1999;16:73–76.

45. Schroeder CG, Pierpaoli PG. Direct order entry by physicians in a computerized hospital information system. *Am J Hosp Pharm.* 1986;43:355–359.

46. Berkowitz LL. Breaking down the barriers: Improving physician buy-in of CPR systems. *Healthcare Inf.* 1997; 14:73–76.

Further Readings

Anderson JG. Information technology in health care: Social and ethical challenges. In: M Witten, ed. *Computational medicine, public health and biotechnology: building a man in the machine.* River Edge, NJ: World Scientific Publishing Co; 1995:1533–1544.

Anderson JG. Clearing the way physician's use of clinical information systems. *Commun ACM.* 1997;40:83–90.

Anderson JG, Aydin CE. Evaluating the impact of health care information systems. *Int J Technol Assess Health Care.* 1997;13:380–393.

Anderson JG, Aydin CE. Overview: theoretical perspectives and methodologies for the evaluation of health care information systems. In: Anderson JG, Aydin CE, Jay SJ, eds. *Evaluating health care information systems: methods and applications.* Thousand Oaks, Calif: Sage; 1994;5–29.

Anderson JG, Aydin CE, Jay SJ, eds. *Evaluating health care information systems: methods and applications.* Thousand Oaks, Calif: Sage Publications; 1994.

Anderson JG, Aydin CE, Kaplan B. An analytical framework for measuring the effectiveness/impacts of computer-based patient record systems. In: Nunamaker JF, Sprague RH Jr, eds. Proceedings: *Twenty-Eighth International conference on Systems Science,* Vol. IV: information systems—collaboration systems and technology, organizational systems and technology. Los Alamitos: IEEE Computer Society Press; 1995:767–776.

Anderson JG, Jay SJ. Hospitals of the future. In: Anderson JG, Jay SJ, eds. *Use and impact of computers in clinical medicine.* New York: Springer-Verlag; 1987;343–350.

Anderson JG, Jay SJ. The social impact of computer technology on physicians. *Computers and Society.* 1991;20:28–33.

Anderson JG, Jay SJ, Schweer HM, Anderson MM. Physician utilization of computers in medical practice: Policy implications based on a structural model. *Soc Sci Med.* 1986;23:259–267.

Anderson JG, Jay SJ, Schweer HM, Anderson MM. Why doctors don't use computers: Some empirical findings. *J Royal Soc Med.* 1986;79:142–144.

Aydin CE. Occupational adaptation to computerized medical information systems. *J Health Soc Behav.* 1989;30:163–179.

Aydin CE. Professional agendas and computer-based patient records: Negotiating for control. *Topics Health Inform Manage.* 1994;15:41–51.

Aydin CE, Anderson JG, Rosen PN, Felitti V, Weng HC. Computers in the consulting room: A case study of clinician and patient perspectives. *Health Care Manage Sci.* 1998;1:61–74.

Berkowitz LL. Breaking down the barriers: Improving physician buy-in of CPR systems. *Healthcare Inf.* 1997;14:73–76.

Brenner DJ, Logan RA. Some considerations in the diffusion of medical technologies: Medical information systems. *Commun Yearbook.* 1980;4:609–623.

Davies BL. A discussion of safety issues for medical robots. In: Taylor RS, Lavallee, S. Burdea GC, Mosges, R, eds. *Computer integrated surgery: technology and clinical applications.* Cambridge, Mass: MIT Press; 1996:287–296.

Davis DA, Taylor-Vaisey A. Translating guidelines into practice: a systematic review of theoretic concepts, practical experience and research evidence in the adoption of clinical practice guidelines. *Can Med Assoc J.* 1997;157:408–416.

Dick RS, Steen EB, Detmer DE, eds. *The computer-based patient record: an essential technology for health care,* Rev. ed. Washington, DC: National Academy Press; 1997.

Esterhay RJ. The medical record: Problem or solution? *MD Comput.* 1993;10:78–80.

Fortess EE, Kapp MB. Medical uncertainty, diagnostic testing, and legal liability. *J Law Med Health Care.* 1985;13:213–218.

Friedman CP, Wyatt JC. *Evaluation methods in medical informatics.* New York: Springer; 1997.

Furfaros C, Muchoney K, Anania-Firouzan P. CPR by the year 2000: a myth? *Healthcare Inform.* 1996;13:45, 47.

Goodman KW. Ethics and system evaluation. *Physic Comput.* 1994;121:12–14.

Gotterbarn D. The moral responsibility of software developers: Three levels of professional software engineering. *J Inform Ethics.* 1995;4:54–64.

Heeks R, Mundy D, Salazar A. Why health care information systems succeed or fail. Information Systems for Public Sector Management Working Paper Series, Working paper No. 9, Institute for Development Policy and Management, University of Manchester, Manchester, UK. Available at: http://www.man.ac.uk/idpm/isps_wp9.htm/. Accessed: August 18, 2000.

Kaplan B. Reducing barriers to physician data entry for computer-based patient records. *Top Health Inform Manage.* 1994;15:24–34.

Kaplan B. Addressing organizational issues into the evaluation of medical systems. *J Am Med Inform Assoc.* 1997;4:94–101.

Kassirer JP. A report card on computer-assisted diagnosis. *N Engl J Med.* 1994;330:1824–1825.

Leveson N, Turner C. An investigation of the Therac-25 accidents. *Computer.* 1993;26:18–41.

Lundsgaarde HP. Evaluating medical expert systems. *Soc Sci Med.* 1987;24:805–819.

Lyytinen K, Hirschheim R. Information systems failure—A survey and classification of empirical literature. *Oxford Surv Inform Technol.* 1987;4:257–309.

Mason RO. Four ethical issues of the information age. *MIS Quarterly.* 1986;10:4–12.

McCormick J. When will smaller medical groups discover computers? *Health Data Manage.* 1997;5:50–63.

Miller RA. Medical diagnostic decision support systems—past, present, and future: A threaded bibliography and brief commentary. *J Am Med Inform Assoc.* 1994;1:8–27.

Miller RA, Schaffner KF, Meisel A. Ethical and legal issues related to the use of computer programs in clinical medicine. *Ann Intern Med.* 1985;102:529–536.

Miller RA, Goodman KW. Ethical challenges in the use of decision-support software. In: Goodman KW, ed. *Ethics, computing and medicine: informatics and the transformation of health care.* New York: Cambridge University Press; 1998:102–115.

Mortimer H. Computer-aided medicine: Present and future issues of liability. *Comput Law J.* 1989;9:177–203.

Neumann PG. *Computer-related risks.* New York: ACM Press; 1995.

Nissenbaum H. Computing and accountability. *Commun ACM.* 1994;37:73–80.

Nissenbaum H. Accountability in a computerized society. *Sci Eng.* 1996;2:25–42.

Schoenbaum SC, Barnett GO. Automated ambulatory medical records systems: An orphan technology. *J Technol Assess Health Care.* 1992;8:598–609.

Siegler M, Toulmin S, Zimring FE, Schaffner KF, eds. *Medical innovation and bad outcomes: legal, social, and ethical responses.* Ann Arbor, Mich: Health Administration Press; 1987.

Simborg DW, Gabler JM. Reengineering the traditional medical record: The view from industry. *MD Computing.* 1992;9:198–200, 272.

Sitting DF, Stead WW. Computer-based physician order entry: The state of the art. *J Am Med Inform Assoc.* 1994;1:108–123.

Snapper JW. Responsibility for computer-based errors. *Metaphilosophy.* 1985;16:289–295.

Stead WW. A quarter-century of computer-based medical records. *MD Computing.* 1989;6:75–81.

Wyatt JC, Spiegelhalter DJ. Evaluating medical expert systems: What to test and how? *Med Inform.* 1990;15:205–217.

Young DW. What makes doctors use computers? Discussion paper. *J Royal Soc Med.* 1984;77:663–667.

7

Online Challenges for Human Subjects Research

"If you need a second opinion, I'll ask my laptop."

Reprinted with permission from Aaron Bacall.

In much the same way that the World Wide Web became indispensable in health, business, entertainment and education, it has become a crucial resource for biomedical researchers and those who participate in their studies. And, as in business, entertainment and education, the Web in biomedical research is no mere source of information, no file cabinet in cyberland. It is a place where people seek out and join research projects and, moreover, as a postmodern agora, the Web is a place where researchers can study the locals and their curious customs.

The use of the Web as a resource—a medium in which researchers describe their projects and the ailing come calling—constitutes what is perhaps the most noteworthy change in biomedical research since the advent of the randomized controlled trial more than a half century ago. The Web as a marketplace—from cancer support-group chat rooms to online discussions of cocaine abuse—is reshaping the very idea of "human subject" and "community" and forcing a reevaluation of what is private and what is public. We should look at these in turn.

Clinical Trials on the Web

On February 29, 2000, the United States Government, under congressional mandate [1], launched a database touted to contain information on nearly all clinical trials for nontrivial illnesses. This database is the work of the National Library of Medicine, whose director, Donald Lindberg [2], described the result as "a single place you can go where the most important information, we hope, will be available to everybody."

The database, ClinicalTrials.gov, is free and contains information about more than 4000 research efforts being conducted in governmental, academic, and drug and device manufacturers' sites.

The impetus for the database was not the government, academia or industry, but patient groups demanding [3] "ready access to information about clinical research studies so that they might be more fully informed about a range of potential treatment options, particularly for very serious diseases."

This is a positive and democratizing enterprise. But there is much at stake when patients, many of whom have life-threatening maladies, search for "potential treatment options." Even what constitutes a "potential treatment option" in a clinical trial is not entirely clear. If the trial demonstrates that an intervention is effective, future treatments incorporating it will be added to the present options, but before an intervention has been demonstrated to be effective, it is not clear that it should be called a treatment. We reserve that term for drugs, devices, and procedures we have reason to believe will work. To suggest otherwise is to go beyond the evidence and perhaps even to engender false hope.

Another consideration for the patient is that if the trial is studying an intervention that will (later) be shown to work, then the trial is worth being on right now, especially if a patient has a serious disease. However, bona fide treatment is needed now, and the trial's results, positive or negative, will not be available until later. In the meantime, the "treatment" being tested might have no positive effect or, indeed, might actually harm individual patients. Clinical trials are, after all, experiments. If it was known that the "treatment" worked, there would be no reason to conduct the experiment; since we do not yet know, it may be irrational to believe that it does.

There is a body of evidence suggesting that patients participating in clinical trials do, in fact, have better outcomes than nonparticipants. This is either because the trials do tend to provide interventions that are better than standard care, or because patients in trials are more closely observed and monitored than those receiving standard care. We know that study subjects are more intensively scrutinized, but this acknowledges they receive a different standard of care. A nonparticipant might reasonably protest that such a discrepancy is unfair.

In the past, patients learned about research studies from their physicians [4], who were trusted to place patient well-being above all other considerations.

The question of how potential trial subjects are best informed of the availability of trials is complicated by the language used to describe the research enterprise. Are patients in trials "subjects" or "participants?" Are they "eligible" (as if for a contest) or do they meet "inclusion criteria?" Did they receive a "treatment" or are they involved in an "experiment?" To describe a study as if it is likely to produce a benefit is unacceptable. Institutional review boards that allow such pitches to appear in consent forms have erred in a fundamental way. (Note the requirement that IRBs scrutinize advertisements soliciting new subjects. To the extent that a Web-based trial listing serves to recruit patients, it might warrant similar scrutiny. An informal review of the descriptions of several studies on ClinicalTrials.gov, however, revealed none of the come-hither language that marks some research solicitations.)

The belief that merely being included in a study will provide a benefit is known as the "therapeutic misconception." In fact, human research participants can be exposed to a broad variety of adverse events, side-effects, and generally untoward outcomes. For example, the drugs tested on cancer patients are powerful and dangerous. Some patients are hurt and some die as subjects in experiments. This is sometimes a cost of biomedical research, but such a cost becomes unacceptable when subjects are desperate, have false beliefs about science, or are inadequately protected by research overseers.

The point here is not that there is anything wrong in using the Web to provide information about clinical trials; on the contrary, an accurate, unbiased and high-quality listing, such as that provided by ClinicalTrials.gov, provides a useful service and a valuable response to demands by patients for more information. However, great care must be taken to protect potential subjects from the therapeutic misconception.

The rules that guide IRBs in protecting subjects must be considered when studies are described on the Web. This kind of attention to ethical detail will help ensure that ClinicalTrials.gov and other sites continue to serve patients, scientists, and society.

Web Users as Subjects

Imagine you are participating in a chat room to discuss your cocaine addiction, eating disorder, or sexual problem. Imagine, further, that one participant or lurker in the chat room does not share the problem but is, instead, a scientist studying the use of chat rooms by addicts, bulimics, or people with sexual dysfunction. You have just become a human subject without your knowledge or consent.

There is a tension here, and it is an interesting and important one. Online behavior is a legitimate focus of scholarly inquiry. There is a lot to be learned about how people use the Web for a variety of health, including mental health, services and benefits. A chat room created for the discussion of intimate problems becomes a prime locus for those who would study the sociological, psychological, and anthropological aspects of these interactions.

Though much can be learned from observing or eavesdropping on these interactions, scientists (and perhaps journalists and others) remain duty-bound to disclose that they are online and gathering information. The problem, well known to psychologists, is that people behave differently if they know they are being observed, and the study is ruined. A scientist who wants to study Web users must then choose between (i) violating rules for confidentiality and informed consent and (ii) self-defeating disclosure.

Do chat-room participants have a bona fide expectation of confidentiality? That depends on a number of factors, including the content and context of communication. We know that the Web has blurred distinctions between public and private domains. Some researchers regard cyberspace as inescapably public [5]. But we also know that if chat-room users were prepared to surrender their privacy, their behavior might not be worth observing.

The solution here should lie in better scientific methods, not in ethical shortcuts. If there is no way to conduct research without violating privacy or confidentiality, then perhaps the research should not be conducted.

That said, there may be a number of cases in which even an expectation of privacy is inadequate to rebut arguments in favor of monitoring, surveillance, and the like. One example is a situation in which it is suspected or feared that people are being harmed or exploited on the Web. There are, in such cases, the beginnings of arguments that greater good is served when confidentiality and consent are temporarily set aside. As a practical matter, the job of ruling on such arguments should probably fall to IRBs, although these overworked and often undersupported panels are frequently ill equipped to evaluate novel research and methods. Those who scrutinize research must broaden their education and improve their ability to do this kind of work.

Case Studies

Clinical Trials

Case 7.1: Information Exchange as a Confounder

Description

An ALS specialist suggested that a few of his patients try Neurontin, a drug that had been approved for epilepsy, and some of his patients shared this information online with other patients. Immediately, other ALS patients began to demand access to Neurontin, though it had not been tested for ALS in clinical trials. Over a few months, as many as 30,000 ALS patients may have taken the new drug. This created problems for investigators running clinical trials on other drugs who feared that some patients enrolled in clinical trials had also begun to take Neurontin. Other patients may have dropped out of the trials altogether fearing they were receiving a placebo.

Source: Bulkeley WM. Untested treatments, cures find stronghold on on-line services. *The Wall Street Journal.* February 27, 1995: A1,A7.

Questions for Discussion

1. Should patients who are enrolled in clinical trials be required to sign a statement promising not to share or discuss their experience with others?
2. Are online support groups for patients with a specific disease potentially detrimental to the conduct of clinical trials of new treatments?

Case 7.2: Web Site with Patient Reports and HIV Treatment

Description

The HIV Treatment Data Project, a collaborative effort among AIDS advocacy organizations, academia, and industry, will provide a Web site containing patients' personal experiences with antiretroviral drugs. The project is intended to provide a repository in which HIV-infected persons record the drugs they are taking, their progress, and how they feel. Supporters of the project say that it will be possible to collect national data on drugs so that clinical decisions can be based on the experiences of a large number of patients. Clinicians and patients must make difficult decisions about complex therapies with limited data from clinical trials. Consequently, the Web site is intended to provide real feedback to patient and health care providers on best practices for treating HIV.

Source: Schoofs M. Click and learn: a new model for AIDS treatment activism. *The Village Voice,* August 19, 1998. Available at: http://www.aegis.com/news/vv/1998/ VV980801.html. Accessed October 14, 2001.

Questions for Discussion

1. The Web makes possible data collection that would otherwise be difficult or impossible. But how can the accuracy of self-reports be assured or evaluated? If it is not clear how to gauge the accuracy of such reports, how useful is this kind of research in the first place? How does self-reporting on the Web differ from self-reporting in other media?
2. How should the confidentiality of individual patients with HIV be protected?

Case 7.3: Promising Trial of New Drug Spurs Demand

Description

Preliminary tests of a new drug to treat chronic myelogenous leukemia (CML) STI-571, resulted in remission in more than 95% of the patients in an early stage of the disease, according to preliminary reports. Information about the success of the drug spread rapidly over the Internet, generating high demand. CML strikes about 10,000 people in Europe and the United States annually. The drug is cytostatic, that is, it is designed to seek out cancer cells without harming healthy ones, as distinguished from less desirable cytotoxic drugs, which kill dividing cells whether or not they are cancerous.

As the number of patients desperately requesting access to the drug increased, Novartis, the company testing the drug, could not meet the unexpected demand. Since the drug was still in an early phase of testing, it had not been approved for general distribution. The company increased production of STI-571, partly in response to demand, and adopted an aggressive strategy for gaining regulatory approval of the drug.

Sources: Moore SD. News about leukemia unexpectedly puts Novartis on the spot. *The Wall Street Journal.* June 6, 2000:A1, A10; and Vastag B. Leukemia drug heralds molecularly targeted era. *J Nat Cancer Inst.* 2000;92:6–8.

Questions for Discussion

1. Preliminary research results often either engender false hope or foreshadow exciting new therapies. How should society respond to rapid, widespread, and unscientific transmission of clinical research data of the sort that seems possible only on the Internet?
2. Free speech issues loom large here, but how would you respond to the claim that drug manufacturers have a duty to discourage lay excitement about potential future remedies? Note that such a duty would not be in the best financial interests of the entities assigned to carry it out, and, in the case of experiments with extremely positive results, may work against the best interest of patients.

The Doctor–Patient Relationship

Case 7.4: Alternative Protocols for Cancer

Description

A woman in Arlington, Virginia, was diagnosed with a brain tumor. Her daughter used the Internet to obtain as much information as possible on the treatment of brain tumors eventually identifying 5 different research protocols for treatment of brain tumors. She then used this information to question why her mother's doctor chose the therapy he did.

Source: Kolata G. Web research transforms visit to the doctor. *The New York Times*. March
  6, 2000: A1, A18.

Questions for Discussion

1. While most reasonable people applaud efforts by patients and family
   members to learn more about health and disease, how should we respond
   to "clinical trial fishing" when used to second-guess a professional's
   treatment plan?
2. Why shouldn't patients and family members be able—even encouraged—
   to question and even challenge their doctors and nurses more?

Case 7.5: Patient–Doctor Conflict over the Appropriateness of a Therapy

Description

A woman with ovarian cancer was being treated at a major medical center. Her
husband went online to examine almost 200 experimental regimens from all over
the world. The couple concluded that intensive therapy with a combination of
drugs would be a more appropriate therapy than the one that she was currently
receiving. Her doctors felt that the current treatment was more appropriate, and
that the woman was too weak to withstand the more intensive treatment regimen.
Source: Brody JE. The health hazards of point-and-click medicine. *The New York Times*.
  August 31, 1999: D6.

Questions for Discussion

1. Should patients who are undergoing therapy for a serious illness be discouraged from investigating alternative therapies over the Internet?
2. Are there any safeguards for patients against faulty, misleading, or erroneous information about alternative therapies?

Case 7.6: Protocol Shopping on the Internet

Description

A 30-year-old woman consulted her physician because she was experiencing
fatigue and back pain. Tests of her blood, urine and bone marrow resulted in a
diagnosis of Stage II multiple myeloma. She consulted several experts in the field
and was recommended to use the standard treatments for multiple myeloma.

However the woman and her husband searched the Internet for all protocols being used to treat multiple myeloma and found a protocol using a high-
dose steroid. She and her husband contacted the IRB at the hospital where she
was being treated to determine if the protocol required board approval before
the drug could be administered. Since additional review of the protocol by
the IRB was deemed unnecessary, the woman insisted on being treated with
the new protocol. This treatment was terminated after 2 cycles due to side

effects. Each week, the woman arrives at the clinic with articles and reports from other clinics about new treatment programs.

Sources: Vaitones V. Protocol "shopping" on the internet. *Cancer Practice.* 1995;3:274–278; and Jadad AR, Gagliardi A. Rating health information on the Internet. *JAMA* 1998;279:611–614.

Questions for Discussion

1. Is the patient's continuous shopping for opinions and protocols counterproductive? Would she be better off to find a physician she trusted and accept his/her recommendations concerning the best treatment for multiple myeloma?
2. Is it important for patients to take an active role in their treatment as this woman has done in order to maintain a sense of control over their lives?
3. Is it difficult for physicians to be responsive to patients while at the same time ensuring that the patient's treatment is scientifically sound?

Research Ethics

Case 7.7: Mining Patient Records

Description

Insurers, drug companies, and health maintenance organizations use medical data to decide which treatments and procedures work and which doctors and health plans are the most effective. For example, the Prudential Insurance Company of America has invested $20 million in a research center to perform outcomes research. Merck & Company, a drug manufacturer, is doing its own outcomes research using data provided by Medco, a company that manages prescription drug benefits for 38 million people in the United States. Medco is owned by Merck & Company and has detailed information on the prescriptions that each doctor writes for each patient and how often the prescription is refilled. They intended to use these data to determine the effectiveness of drugs.

This research center had begun to gather information from medical records, medical diagnoses, insurance forms, and prescriptions. For example, information from patients who took cholesterol-lowering drugs could be used to determine whether these patients experienced fewer heart attacks than patients with similar levels of cholesterol who did not take the drugs. Health maintenance organizations are using these data to create profiles on physicians permitting the identification of physicians who use too many or two few medical resources in treating patients. For example, Healthsource Inc, an HMO based in Hooksett, NH, with 850,000 members in 12 states, compares each physician to the average every quarter. The company punishes physicians who perform poorly by withholding a percentage of their income or even discontinuing their contracts with the HMO.

Source: Kolata G. The healthcare debate: finding what works. *The New York Times*. August
  9, 1994:A20.

Questions for Discussion

1. Are data from medical records, insurance claims forms, and pharmacy
   records too biased to be used to determine the relative effectiveness of
   drugs and medical procedures?
2. Should outcomes research centers that use data from individuals' medical and pharmacy records be required to obtain patients' approval before
   using these data?

Case 7.8: Snooping on Internet Chat Rooms

Description

Researchers are secretly recording and studying messages posted on Internet
chat rooms and newsgroups. Participants in the chat rooms and news groups
often disclose sensitive information about illnesses, addictions, personal
circumstances, sexual behavior and abuse, and they are frequently unaware
that information that they provide online is not confidential.

  Some of this research has been published with quotations from people who
participated in the chat rooms. In one instance, a researcher presented findings at a psychological association conference from a study of persons participating in an eating disorder bulletin board. A woman in the audience who
had contributed to the bulletin board was upset and said she felt violated.
Source: Francescani C. Cybersnoop docs spy on your E-chats. *New York Post*, May 1,
  2000: 22.

Questions for Discussion

1. Is research that involves secretly observing and recording information
   exchanged on an Internet bulletin board or chat room different from observing and recording human behavior in other public settings?
2. Is it unethical to conduct this type of research without first securing
   consent from all participants?
3. Should research protocols that involve collecting information from
   Internet chat rooms and bulletin boards have to be approved by IRBs?

Case 7.9: Tracing Syphilis in Cyberspace

Description

A syphilis outbreak among gay men who met sex partners through an Internet
chat room has challenged existing models of partner notification and
community education. Privacy protections prevented public health
officials from learning the identities of the sex partners; indeed, the

Internet service provider that hosted the chat room told the San Francisco Department of Public Health that it would not release identifying information without a federal subpoena. As a result, partner notification was limited to screen names. Public health staffers sent e-mail messages to the screen names they were able to acquire, and 42% of the named partners were notified and tested. The mean number of sexual partners medically evaluated per index case was 5.9.

Source: Klausner JD, Wolf W, Fischer-Ponce L, Zolt I, Katz MH. Tracing a syphilis outbreak through cyberspace. *JAMA*. 2000;284:447–449.

Questions for discussion

1. It is clear the Internet can facilitate risky health behaviors. To what extent do public health concerns justify increased monitoring of the Internet (in much the same way as other venues for the spread of contagion are scrutinized)?

2. How should society weigh the expectation of online privacy and confidentiality against the needs of epidemiology and public health?

Note: Parts of this chapter are adapted from Goodman KW. Using the Web as a research tool. MD Comput. 2000;17:13–14, with permission.

References

1. FDA Modernization Act of 1997, Public Law 105–115, 105th Congress. Section 113, Information program on clinical trials for serious or life-threatening diseases. Food and Drug Administration Web site. Available at: http://www.fda.gov/cder/guidance/105-115.htm. Accessed June 13, 2000.
2. Associated Press. Clinical trials database for patients goes online. *The New York Times*. February 29, 2000: A4.
3. McCray AT, Ide NC. Design and implementation of a national clinical trials registry. *J Am Med Inf Assoc*. 2000;7:313–323.
4. Kahn JP. Shopping for clinical trials: New NIH Web site raises ethical issues. CNN.com Web site, March 6, 2000. Available at: http://www.cnn.com/2000/HEALTH/03/06/ethics.matters/index.htm. Accessed June 12, 2000.
5. Siang S. Researching ethically with human subjects in cyberspace. *Professional Ethics Report*. 2000;12:1,7–8.

Further Readings

Dresser R. Surfing for studies: clinical trials on the Internet. *Hastings Center Report*. November-December 1999:26–27.
Eysenbach G, Diepgen TL. Epidemiological data can be gathered with World Wide Web. *BMJ*. 1998;316:72.
Houston JD, Fiore DC. Online medical surveys: using the Internet as a research tool. *MD Comput*. 1998;15:116–120.

Lakeman R. Using the Internet for data collection in nursing research. *Comput Nurs.* 1997;15:269–275.

Mueller J. Research on-line: human participants ethics issues. University of Calgary Department of Psychology Web site, Available at: www.psych.ucalgary.ca/Research/Ethics/online.html. Accessed Sept. 26, 2000.

Schleyer TKL, Forrest JL. Methods for the design and administration of Web-based surveys. *J Am Med Inf Assoc.* 2000;7:416–425.

Thomas B, Stamier LL, Lafreniere K, Dumata R. The Internet: An effective tool for nursing research with women. *Comput Nurs.* 2000;18:13–18.

Toomey KE, Rothenberg RB. Sex and cyberspace—virtual networks leading to high-risk sex. *JAMA.* 2000;284:485–487.

Turner JL, Turner DB. Using the Internet to perform survey research. *Syllabus.* 1998;12:58–61.

Wyatt J. When to use Web-based surveys. *J Am Med Inf Assoc.* 2000;7:426–430.

Appendices

Ethical Standards
for Health Web Sites

A number of organizations are attempting to develop ethical standards for health and medical Web sites. Some of these are the HONcode of Conduct, published by the Health on the Net Foundation located in Geneva, Switzerland; the eHealth Code of Ethics, produced by the Internet Healthcare Coalition; the Model Privacy Statement by TRUSTe, an independent nonprofit organization located in Cupertino, California; Ethical Principles for Offering Health Services to Consumers, published by Hi-Ethics or the Health Internet Ethics, a group of 20 organizations with health-related Web sites; Principles Governing AMA Publications Web Sites, developed by the American Medical Association; and the Criteria for Assessing the Quality of Health Information on the Internet and the Information Quality Tool to evaluate Web sites, published by the Health Information Technology Institute of Mitretek Systems. All of these standards are reproduced with permission in the Appendices.

Appendix 1

HON Code of Conduct (HONcode) for Medical and Health Web Sites

(Available at: http://www.hon.ch/Honcode/conduct.html/. Accessed: October 15, 2001)

Principles

1. **Authority:** Any medical or health advice provided and hosted on this site will only be given by medically trained and qualified professionals unless a clear statement is made that a piece of advice offered is from a non-medically qualified individual or organisation.
2. **Complementarity:** The information provided on this site is designed to support, not replace, the relationship that exists between a patient/site visitor and his/her existing physician.
3. **Confidentiality:** Confidentiality of data relating to individual patients and visitors to a medical/health Web site, including their identity, is respected by this Web site. The Web site owners undertake to honour or exceed the legal requirements of medical/health information privacy that apply in the country and state where the Web site and mirror sites are located.
4. **Attribution:** Where appropriate, information contained on this site will be supported by clear references to source data and, where possible, have specific HTML links to that data. The date when a clinical page was last modified will be clearly displayed (e.g. at the bottom of the page).
5. **Justifiability:** Any claims relating to the benefits/performance of a specific treatment, commercial product or service will be supported by appropriate, balanced evidence in the manner outlined above in Principle 4.
6. **Transparency of authorship:** The designers of this Web site will seek to provide information in the clearest possible manner and provide contact addresses for visitors that seek further information or support. The

Webmaster will display his/her E-mail address clearly throughout the Web site.

7. **Transparency of sponsorship:** Support for this Web site will be clearly identified, including the identities of commercial and non-commercial organisations that have contributed funding, services or material for the site.

8. **Honesty in advertising & editorial policy:** If advertising is a source of funding it will be clearly stated. A brief description of the advertising policy adopted by the Web site owners will be displayed on the site. Advertising and other promotional material will be presented to viewers in a manner and context that facilitates differentiation between it and the original material created by the institution operating the site.

Appendix 2

eHealth Code of Ethics

(Available at: http://www.ihealthcoalition.org/ethics/code0524.pdf. Accessed: October 15, 2001)

Vision Statement

The goal of the eHealth Code of Ethics is to ensure that people worldwide can confidently and with full understanding of known risks realize the potential of the Internet in managing their own health and the health of those in their care.

Introduction

The Internet is changing how people give and receive health information and health care. All people who use the Internet for health-related purposes— patients, health care professionals and administrators, researchers, those who create or sell health products or services, and other stakeholders—must join together to create a safe environment and enhance the value of the Internet for meeting health care needs.

Because health information, products, and services have the potential both to improve health and to do harm, organizations and individuals that provide health information on the Internet have obligations to be trustworthy, provide high quality content, protect users' privacy, and adhere to standards of best practices for online commerce and online professional services in health care.

People who use Internet health sites and services share a responsibility to help assure the value and integrity of the health Internet by exercising judgment in using sites, products, and services, and by providing meaningful feedback about online health information, products, and services.

Definitions

Health information includes information for staying well, preventing and managing disease, and making other decisions related to health and health care.

It includes information for making decisions about health products and health services.
It may be in the form of data, text, audio, and/or video.
It may involve enhancements through programming and interactivity.

Health products include drugs, medical devices, and other goods used to diagnose and treat illnesses or injuries or to maintain health. Health products include both drugs and medical devices subject to regulatory approval by agencies such as the U.S. Food and Drug Administration or U.K. Medicines Control Agency *and* vitamin, herbal, or other nutritional supplements and other products not subject to such regulatory oversight.

Health services include specific, personal medical care or advice; management of medical records; communication between health care providers and/or patients and health plans or insurers, or health care facilities regarding treatment decisions, claims, billing for services, etc.; and other services provided to support health care.

Health services also include listserves, bulletin boards, chat rooms, and other online venues for the exchange of health information.

Like health information, health services may be in the form of data, text, audio, and/or video, and may involve enhancements through programming and interactivity.

Anyone who uses the Internet for health-related reasons has a right to expect that organisations and individuals who provide health information, products or services online will uphold the following guiding principles:

Candor

People who use the Internet for health-related purposes need to be able to judge for themselves that the sites they visit and services they use are credible and trustworthy. Sites should clearly indicate

- who owns or has a significant financial interest in the site or service
- what the purpose of the site or service is

For example, whether it is solely educational, sells health products or services, or offers personal medical care or advice

- any relationship (financial, professional, personal, or other) that a reasonable person would believe would likely influence his or her perception of the information, products, or services offered by the site

    For example, if the site has commercial sponsors or partners, who those sponsors/partners are and whether they provide content for the site

Honesty

People who seek health information on the Internet need to know that products or services are described truthfully and that information they receive is not presented in a misleading way. Sites should be forthright

- in all content used to promote the sale of health products or services
- in any claims about the efficacy, performance, or benefits of products or services

They should clearly distinguish content intended to promote or sell a product, service, or organization from educational or scientific content.

Quality

To make wise decisions about their health care, people need and have the right to expect that sites will provide accurate, well-supported information and products and services of high quality.

To assure that the health information they provide is accurate, eHealth sites and services should make good faith efforts to

- evaluate information rigorously and fairly, including information used to describe products or services
- provide information that is consistent with the best available evidence
- assure that when personalized medical care or advice is provided that care or advice is given by a qualified practitioner
- indicate clearly whether information is based on scientific studies, expert consensus, or professional or personal experience or opinion
- acknowledge that some issues are controversial and when that is the case make good faith efforts to present all reasonable sides in a fair and balanced way

    For example, advise users that there are alternative treatments for a particular health condition, such as surgery or radiation for prostate cancer

Information and services must be easy for consumers to understand and use. Sites should present information and describe products or services

- in language that is clear, easy to read, and appropriate for intended users

  For example, in culturally appropriate ways in the primary language (or languages) of the site's expected and

- in a way that accommodates special needs users may have

  For example, in large type or through audio channels for users whose vision is impaired

Sites that provide information primarily for educational or scientific purposes should guarantee the independence of their editorial policy and practices by assuring that only the site's content editors determine editorial content and have the authority to reject advertising that they believe is inappropriate.

Consumers have a right to expect that the information they receive is up to date. Sites should clearly indicate

- when the site published the information it provides (and what version of the information users are seeing if it has been revised since it was first published)
- when the site most recently reviewed the information
- whether the site has made substantive changes in the information and if so, when the information was most recently updated

Individuals need to be able to judge for themselves the quality of the health information they find on the Internet. Sites should describe clearly and accurately how content is developed for the site by telling users

- what sources the site or content provider has used, with references or links to those sources
- how the site evaluates content and what criteria are used to evaluate content, including on what basis the site decides to provide specific links to other sites or services

  For example, by describing the site's editorial board and policies

When health products or services are subject to government regulation, sites should tell users whether those products (such as drugs or medical devices) have been approved by appropriate regulatory agencies, such as the U.S. Food and Drug Administration or U.K. Medicines Control Agency.

Informed Consent

People who use the Internet for health-related reasons have the right to be informed that personal data may be gathered, and to choose whether they will allow their personal data to be collected and whether they will allow it to be used or shared. And they have a right to be able to choose, consent, and control when and how they actively engage in a commercial relationship. Sites should clearly disclose

- that there are potential risks to users' privacy on the Internet

  For example, that other organisations or individuals may be able to collect personal data when someone visits a site, without that site's knowledge; or that some jurisdictions (such as the European Union) protect privacy more stringently than others

Sites should not collect, use, or share personal data without the user's *specific affirmative consent*. To assure that users understand and make informed decisions about providing personal data, sites should indicate clearly and accurately

- what data is being collected when users visit the site

  For example, data about which parts of the site the user visited, or the user's name and email address, or specific data about the user's health or online purchases

- who is collecting that data

  For example, the site itself, or a third party

- how the site will use that data

  For example, to help the site provide better services to users, as part of a scientific study, or to provide personalised medical care or advice

- whether the site knowingly shares data with other organisations or individuals and if so, what data it shares
- which organisations or individuals the site shares data with and how it expects its affiliates to use that data

  For example, whether the site will share users' personal data with other organisations or individuals and for what purposes, and note when personal data will be shared with organizations or individuals in other countries

- obtain users affirmative consent to collect, use, or share personal data in the ways described

  For example, to collect and use the visitor's personal data in scientific research, or for commercial reasons such as sending information about new products or services to the user, or to share his or her personal data with other organisations or individuals

- what consequences there may be when a visitor refuses to give personal data

For example, that the site may not be able to tailor the information it provides to the visitor's particular needs, or that the visitor may not have access to all areas of the site

"E-commerce" sites have an obligation to make clear to users when they are about to engage in a commercial transaction and to obtain users' specific affirmative consent to participate in that commercial transaction.

Privacy

People who use the Internet for health-related reasons have the right to expect that personal data they provide will be kept confidential. Personal health data in particular may be very sensitive, and the consequences of inappropriate disclosure can be grave. To protect users, sites that collect personal data should

* take reasonable steps to prevent unauthorised access to or use of personal data

  For example, by "encrypting" data, protecting files with passwords, or using appropriate security software for all transactions involving users' personal medical or financial data

* make it easy for users to review personal data they have given and to update it or correct it when appropriate
* adopt reasonable mechanisms to trace how personal data is used

  For example, by using "audit trails" that show who viewed the data and when

* tell how the site stores users' personal data and for how long it stores that data
* assure that when personal data is "de-identified" (that is, when the user's name, email address, or other data that might identify him or her has been removed from the file) it cannot be linked back to the user

Professionalism in Online Health Care

Physicians, nurses, pharmacists, therapists, and all other health care professionals who provide specific, personal medical care or advice online should

* abide by the ethical codes that govern their professions as practitioners in face-to-face relationships
* do no harm
* put patients' and clients' interests first
* protect patients' confidentiality

- clearly disclose any sponsorships, financial incentives, or other information that would likely affect the patient's or client's perception of professional's role or the services offered
- clearly disclose what fees, if any, will be charged for the online consultation and how payment for services is to be made
- obey the laws and regulations of relevant jurisdiction(s), including applicable laws governing professional licensing and prescribing

The Internet can be a powerful tool for helping to meet patients' health care needs, but users need to understand that it also has limitations. Health care professionals who practice on the Internet should clearly and accurately

- identify themselves and tell patients or clients where they practice and what their professional credentials are
- describe the terms and conditions of the particular online interaction

  For example, whether the health care professional will provide general advice about a particular health condition or will make specific recommendations and or referrals for the patient or client, or whether the health care professional can and will or cannot and will not prescribe medications in the particular situation

- make good faith efforts to understand the patient's or client's particular circumstances and to help him or her identify health care resources that are available locally

  For example, to help the patient or client determine whether particular treatment is available in his or her home community or only from providers outside his or her community

- give clear instructions for follow-up care when appropriate or necessary

Health care professionals who offer personal medical services or advice online should

- clearly and accurately describe the constraints of online diagnosis and treatment recommendations

  For example, providers should stress that because the online health care professional cannot examine the patient, it is important for patients to describe their health care needs as clearly they can

- help "e-patients" understand when online consultation can and when it cannot and should not take the place of a face-to-face interaction with a health care provider

Responsible Partnering

People need to be confident that organisations and individuals who operate on the Internet undertake to partner only with trustworthy individuals or organisations. Whether they are for-profit or nonprofit, sites should

- make reasonable efforts to ensure that sponsors, partners, or other affiliates abide by applicable law and uphold the same ethical standards as the sites themselves
- insist that current or prospective sponsors not influence the way search results are displayed for specific information on key words or topics

And they should indicate clearly to users

- whether links to other sites are provided for information only or are endorsements of those other sites
- when they are leaving the site

    For example, by use of transition screens

Accountability

People need to be confident that organisations and individuals that provide health information, products, or services on the Internet take users' concerns seriously and that sites make good faith efforts to ensure that their practices are ethically sound. eHealth sites should

- indicate clearly to users how they can contact the owner of the site or service and/or the party responsible for managing the site or service

    For example, how to contact specific manager(s) or customer service representatives with authority to address problems

- provide easy-to-use tools for visitors to give feedback about the site and the quality of its information, products, or services
- review complaints from users promptly and respond in a timely and appropriate manner

Sites should encourage users to notify the site's manager(s) or customer service representatives if they believe that a site's commercial or noncommercial partners or affiliates, including sites to which links are provided, may violate law or ethical principles.

eHealth sites should describe their policies for self-monitoring clearly for users, and should encourage creative problem solving among site staff and affiliates.

Appendix 3

TRUSTe's Consumer Privacy Protection Guidelines

(Available at: http://www.truste.org/education/protection_guidelines.html. Accessed: October 15, 2001)

1. **Read Privacy Statements.** Look for the Web site's privacy statement and read it thoroughly. Steer clear of Web sites that don't have a privacy statement. A privacy statement is a legally binding document that describes the personal information gathering and dissemination practices of a Web site. Make sure you understand how your information will be used before you do business with a Web site.

2. **Seal Programs.** Look for approval seals indicating that the privacy policies of the site are being monitored by an outside agency, such as TRUSTe. These programs allow you to turn to a third party if you feel that your privacy has been violated. Click on the TRUSTe Privacy Seal to see what information about you is gathered and with whom it is shared, as well as how to prevent the sharing of your personal information and how to correct inaccuracies.

3. **Credit cards.** The same consumer protection laws that apply in the mall apply on the Internet. Using credit cards allows you to contest any charges if the merchandise does not live up to the promotion. In addition, federal law limits your liability to $50 for purchases made with stolen credit card information.

4. **Security.** While no Web site is hack proof, you should only place credit card orders through secure servers. Most Web merchants alert you when you are entering their secure servers. Also check that the URL (Web address) begins with "https" rather than "http"; this indicates that you have entered the secure area. When using newer browsers you will also see either a closed lock or a solid key symbol in the status bar at the bottom of your screen.

5. **Common sense.** Don't disclose information you wouldn't want to disclose over the phone or in person. Remember: you can always contact the Web site to find out more about its privacy and security practices before you make a purchase.

6. **Protect children.** Just as you would teach your kids street smarts, help your children be "cybersmart" by giving them guidance on what to look out for when surfing the Net. Keep in mind that Web sites directed at children under 13 are required by law to adhere to certain privacy practices. These practices include obtaining verifiable parental consent before children participate in certain activities. Look for the TRUSTe Children's Privacy Seal, a seal with expanded safeguards, on child oriented Web sites.

Appendix 4

TRUSTe Model Privacy Statement

This confirms that [COMPANY X] is a licensee of the TRUSTe Privacy Program. This privacy statement discloses the privacy practices for [URL of COMPANY X WEBSITE].

TRUSTe is an independent, non-profit organization whose mission is to build users' trust and confidence in the Internet by promoting the use of fair information practices. Because this web site wants to demonstrate its commitment to your privacy, it has agreed to disclose its information practices and have its privacy practices reviewed for compliance by TRUSTe. By displaying the TRUSTe trustmark, this web site has agreed to notify you of:

1. What personally identifiable information of yours or third party personally identification is collected from you through the web site?
2. The organization collecting the information
3. How the information is used
4. With whom the information may be shared
5. What choices are available to you regarding collection, use and distribution of the information?
6. The kind of security procedures that are in place to protect the loss, misuse or alteration of information under [NAME OF COMPANY] control
7. How you can correct any inaccuracies in the information.

If you feel that this company is not abiding by its posted privacy policy, you should first contact [INSERT NAME OF INDIVIDUAL, DEPARTMENT OR GROUP RESPONSIBLE FOR INQUIRIES] by [INSERT CONTACT INFORMATION; EMAIL, PHONE, POSTAL MAIL, ETC.] If you do not receive acknowledgment of your inquiry or your inquiry has not been satisfactorily addressed, you should then contact TRUSTe at http://www.truste.org. TRUSTe will then serve as a liaison with the Web site to resolve your concerns.

Information Collection and Use

Company X is the sole owner of the information collected on this site. We will not sell, share, or rent this information to others in ways different from what is disclosed in this statement. Company X collects information from our users at several different points on our website.

Registration

In order to use this website, a user must first complete the registration form. During registration a user is required to give their contact information (such as name and email address). This information is used to contact the user about the services on our site for which they have expressed interest. It is optional for the user to provide demographic information (such as income level and gender), and unique identifiers (such as social security number), but encouraged so we can provide a more personalized experience on our site.

Order

We request information from the user on our order form. Here a user must provide contact information (like name and shipping address) and financial information (like credit card number, expiration date). This information is used for billing purposes and to fill customer's orders. If we have trouble processing an order, this contact information is used to get in touch with the user.

Cookies

A cookie is a piece of data stored on the user's hard drive containing information about the user. Usage of a cookie is in no way linked to any personally identifiable information while on our site. Once the user closes their browser, the cookie simply terminates. For instance, by setting a cookie on our site, the user would not have to log in a password more than once, thereby saving time while on our site. If a user rejects the cookie, they may still use our site. The only drawback to this is that the user will be limited in some areas of our site. For example, the user will not be able to participate in any of our Sweepstakes, Contests or monthly Drawings that take place. Cookies can also enable us to track and target the interests of our users to enhance the experience on our site.

Some of our business partners use cookies on our site (for example, advertisers). However, we have no access to or control over these cookies.

Log Files

We use IP addresses to analyze trends, administer the site, track user's movement, and gather broad demographic information for aggregate use. IP addresses are not linked to personally identifiable information.

Sharing

We will share aggregated demographic information with our partners and advertisers. This is not linked to any personal information that can identify any individual person.

We use an outside shipping company to ship orders, and a credit card processing company to bill users for goods and services. These companies do not retain, share, store or use personally identifiable information for any secondary purposes.

We partner with another party to provide specific services. When the user signs up for these services, we will share names, or other contact information that is necessary for the third party to provide these services.

These parties are not allowed to use personally identifiable information except for the purpose of providing these services.

Links

This web site contains links to other sites. Please be aware that we [COMPANY X] are not responsible for the privacy practices of such other sites. We encourage our users to be aware when they leave our site and to read the privacy statements of each and every web site that collects personally identifiable information. This privacy statement applies solely to information collected by this Web site.

Newsletter

If a user wishes to subscribe to our newsletter, we ask for contact information such as name and email address.

Surveys & Contests

From time-to-time our site requests information from users via surveys or contests. Participation in these surveys or contests is completely voluntary and the user therefore has a choice whether or not to disclose this information. Information requested may include contact information (such as name and shipping address), and demographic information (such as zip code, age level). Contact information will be used to notify the winners and award prizes. Survey information will be used for purposes of monitoring or improving the use and satisfaction of this site.

Tell-A-Friend

If a user elects to use our referral service for informing a friend about our site, we ask them for the friend's name and email address. [COMPANY X] will automatically send the friend a one-time email inviting them to visit the site. [COMPANY X] stores this information for the sole purpose of sending this one-time email. The friend may contact [COMPANY X] at [INSERT URL] to request the removal of this information from their database.

Security

This website takes every precaution to protect our users' information. When users submit sensitive information via the website, your information is protected both online and offline.

When our registration/order form asks users to enter sensitive information (such as credit card number and/or social security number), that information is encrypted and is protected with the best encryption software in the industry — SSL. While on a secure page, such as our order form, the lock icon on the bottom of Web browsers such as Netscape Navigator and Microsoft Internet Explorer becomes locked, as opposed to unlocked, or open, when you are just 'surfing'. To learn more about SSL, follow this link [INSERT LINK].

While we use SSL encryption to protect sensitive information online, we also do everything in our power to protect user-information offline. All of our users' information, not just the sensitive information mentioned above, is restricted in our offices. Only employees who need the information to perform a specific job (for example, our billing clerk or a customer service representative) are granted access to personally identifiable information. Our employees must use password-protected screen-savers when they leave their desk. When they return, they must re-enter their password to re-gain access to your information. Furthermore, ALL employees are kept up-to-date on our security and privacy practices. Every quarter, as well as any time new policies are added, our employees are notified and/or reminded about the importance we place on privacy, and what they can do to ensure our customers' information is protected. Finally, the servers that we store personally identifiable information on are kept in a secure environment, behind a locked cage.

If you have any questions about the security at our website, you can send an email to security@thiswebsite.com.

Supplementation of Information

In order for this website to properly fulfill its obligation to our customers, it is necessary for us to supplement the information we receive with information from 3rd party sources. For example, to determine if our customers qualify for one of our credit cards, we use their name and social security number to request a credit report. Once we determine a user's credit-worthiness, this document is destroyed.

(or)

In order for this website to enhance its ability to tailor the site to an individual's preference, we combine information about the purchasing habits of users with similar information from our partners, Company Y & Company Z, to create a personalized user profile. When a user makes a purchase from either of these two companies, the companies collect and share that purchase information with us so we can tailor the site to our users' preferences.

Special Offers

We send all new members a welcoming email to verify password and username. Established members will occasionally receive information on products, services, special deals, and a newsletter. Out of respect for the privacy of our users we present the option to not receive these types of communications. Please see our choice and opt-out below.

Site and Service Updates

We also send the user site and service announcement updates. Members are not able to un-subscribe from service announcements, which contain important information about the service. We communicate with the user to provide requested services and in regards to issues relating to their account via email or phone.

Correction/Updating Personal Information:

If a user's personally identifiable information changes (such as your zip code), or if a user no longer desires our service, we will endeavor to provide a way to correct, update or remove that user's personal data provided to us. This can usually be done at the member information page or by emailing our Customer Support. [Some sites may also provide telephone or postal mail options for updating or correcting personal information].

Choice/Opt-out

Our users are given the opportunity to 'opt-out' of having their information used for purposes not directly related to our site at the point where we ask for the information. For example, our order form has an 'opt-out' mechanism so users who buy a product from us, but don't want any marketing material, can keep their email address off of our lists.

Users who no longer wish to receive our newsletter or promotional materials from our partners may opt-out of receiving these communications by replying to unsubscribe in the subject line in the email or email us at support@thiswebsite.com [Some sites are able to offer opt-out mechanisms on member information pages and also supply a telephone or postal option as a way to opt-out.]

Users of our site are always notified when their information is being collected by any outside parties. We do this so our users can make an informed choice as to whether they should proceed with services that require an outside party, or not.

Notification of Changes

If we decide to change our privacy policy, we will post those changes on our Homepage so our users are always aware of what information we collect, how we use it, and under circumstances, if any, we disclose it. If at any point we decide to use personally identifiable information in a manner different from that stated at the time it was collected, we will notify users by way of an email. Users will have a choice as to whether or not we use their information in this different manner. We will use information in accordance with the privacy policy under which the information was collected.

Appendix 5

Health Internet Ethics: Ethical Principles For Offering Internet Health Services to Consumers

(Available: http://www.hiethics.com/ Accessed: October 15, 2001)

As Hi-Ethics members, we are committed to ensuring that individual consumers can realize the full benefits of the Internet to improve their health and that of their families. To fulfill our commitment, we are dedicated to meeting the following goals:

Internet health services that reflect high quality and ethical standards;
Providing health information that is trustworthy and up-to-date;
Keeping personal information private and secure, and employing special precautions for any personal health information; and
Empowering consumers to distinguish online health services that follow our principles from those that do not.

Informed by these goals, we adopt the following ethical principles. We believe that in living by these principles, we can improve the consumer's experience with online health information and services. We have provided a glossary of terms with special meanings at the end of this document.

1. Privacy Policies

Our members will adopt a privacy policy that is easy for consumers to find, read, and understand. Our privacy policies will—

A. Provide users with reasonable notice of our information practices, including disclosure of—

1. collection or use of any information about the user;
2. collection or use of aggregate data; and
3. what, if any, access to personal information collected on our health web site we provide to unrelated third parties.

B. Provide consumers with a meaningful choice on our health web site to accept or decline our proposed collection and use of personal information provided by the consumer including, if any, consent to the transfer of information to third parties.

C. Contain a positive commitment from us to use security procedures to protect personal information we collect from misuse.

D. Provide, where appropriate, procedures for consumers to review and correct their personal information that we maintain, or to request that we delete the information, and include a description of the effect of any changes on other information about the user that we maintain.

2. Enhanced Privacy Protection for Health-Related Personal Information

A. If we collect health-related personal information, we will only use it for the purposes for which a reasonable consumer would expect us to use it or as agreed to by the consumer.

B. We will not disclose health-related personal information to an unrelated third party and/or for unrelated purposes without first obtaining the consent of the consumer (by means of an explicit "opt-in" procedure).

C. When we make significant changes to our privacy policies that affect the use of the health-related personal information we collect, we will give notice to our users. We will not make use of information we gathered from individuals prior to a significant change in polic without first obtaining their consent for any new uses. We may also make non-significant changes to our privacy policies that will not affect our use of a consumer's personal information. We will post such changes on our health web site.

3. Safeguarding Consumer Privacy in Relationships with Third Parties

A. Where third parties have access to health-related personal information from our site, our agreements with these third parties will follow these principles in giving consumers notice and choice with respect to that third party's access and use.

B. Where we have relationships with third parties, we will adopt procedures to tell consumers if third parties have access to personal information about them from our site.

C. We will take appropriate precautions to prevent inadvertent disclosures of personal information to third parties and will take immediate steps to eliminate such disclosures, if they occur, once they have come to our attention.

D. We will not allow third parties any access to non-personal individual information collected on our site unless the third party agrees that it will not use the information to identify individuals.

4. Disclosure of Ownership and Financial Sponsorship

We will disclose those who have major financial interests in us or the health web sites we operate, and those who give us significant funding or other assistance. We will—

A. Clearly state who owns any health web site we operate.
B. Clearly identify those who hold an ownership interest of 10% or more in our company, and those whose financial contributions to our health web site represent 10% or more of the annual revenues of our company. Financial contributions means both cash and in-kind services or materials by persons who are not otherwise identified as sponsors.

5. Identifying Advertising and Health Information Content Sponsored by Third Parties

A. We will clearly distinguish advertising from health information content, using identifying words, design, or placement. We will design our health web sites to avoid confusion between advertising and health information content.
B. We will clearly disclose significant relationships between commercial sponsors and our health information content by identifying a sponsor's involvement in—

   1. selecting or preparing health information content that appears on our health web site, including any sponsorship of priority listings in search engine results, product listings, or other preferences in presentation of information to consumers; and
   2. any "co-branding" of health information content or Internet health services.

C. We will provide consumers with a policy that is easy for consumers to find, read and understand regarding our acceptance of advertising and of health information content sponsored by others. Our policy will disclose—

   1. how we identify advertising and commercially sponsored health information content on our health web site;
   2. how we may obtain revenues from third parties related to advertising and health information content sponsored by others on our health web site, including advertising revenues, commissions on consumer purchases, fees based on consumer use of links to other web sites,

and revenues for transfer or use of information about users, including aggregate data;

3. whether we target advertising or sponsored health information content to consumers based on information about them or their use of our health web site; and

4. whether we intend any links to other web sites, logos, or marks of other companies, or any co-branding to constitute recommendations to the consumer.

6. Promotional Offers, Rebates, and Free Items or Services

We will comply with existing federal and state laws regarding any promotions, rebates, and free or discounted offers on our health web sites.

7. Quality of Health Information Content

A. We will not make claims of therapeutic benefit without reasonable support, or deliberately provide false or misleading information.

B. We will not accept advertising or sponsored health information content that we know either contains false or misleading claims or promotes ineffective or dangerous products.

C. We will have an editorial policy that is easy for consumers to find, read, and understand. Our editorial policy will describe procedures we use for evaluating the quality of the health information content on our health web site, whether created by us or obtained from others.

8. Authorship and Accountability

A. We will disclose any cases where we have placed health information content on our health web site because of sponsorship or other support from a third party. In addition to identifying the sponsor, we will clearly disclose significant relationships between the commercial sponsor and our health information content by identifying the sponsor's involvement with that content.

B. Where we reproduce health information content created by third parties, we will clearly disclose the author and/or source of the material and the date of the material or its last update.

C. Where we present health information content as the result of clinical experience or scholarly research, we will clearly disclose the actual author(s) of the health information content.

D. Where we create health information content for use by consumers, we will provide consumers general information about our authors and their qualifications, our editorial policy, and, if any, our expert review process.

E. Where we create health information content, we will clearly disclose the date it was created or last updated.

F. We will have a conflict of interest policy for all authors that is easy for consumers to find, read, and understand. We will disclose all affiliations and financial relationships of authors consistent with our policy.

9. Disclosure of Source and Validation for Self-Assessment Tools

A. Where we offer self-assessment tools, we will disclose their source and appropriately describe the scientific basis for their operation.

B. We will also describe how we maintain self-assessment tools, including a description of any formal evaluation process and the date of the last review or update.

10. Professionalism

A. We believe that current codes of ethics apply when health care professionals use health web sites to provide professional care. However, these codes do not apply to every interaction between a consumer and a professional. Our health web sites shall provide conspicuous and appropriate information for consumers to understand when they are and are not in an interaction with a health professional that is covered by the ethical standards of the profession.

B. Where we allow health care professionals to engage in professional care on our health web sites, we will design Internet health services to enable health care professionals to adhere to professional ethical principles in the online environment. We will continue to evolve new standards of practice to meet the changing expectations created by consumers' use of Internet health services.

C. Internet health services directed to and for use by health care professionals are beyond the scope of these principles.

11. Qualifications

A. We will provide the credentials and qualifications of persons responsible for health care services delivered via our consumer health web sites. If applicable, we will also provide information about professional licensure.

B. We will disclose whether we verify information regarding health care professionals or others who provide services or information on our health web sites.

12. Transparency of Interactions, Candor and Trustworthiness

A. We will inform consumers who use our Internet health services of the risks, responsibilities, and reasonable expectations associated with their use of our services. We will make sure that this information is easy for consumers to find, read, and understand.

B. We will strive to make it apparent to consumers when they move within a site, or leave one site for another, and when the move changes the risks, responsibilities, and expectations associated with their activities.

13. Disclosure of Limitations

We will advise consumers of any limitations of our health web site as a source of health care services. In particular, we will state that online health services and health information content cannot replace a health professional-patient relationship, and that consumers should always consult with a professional for diagnosis and treatment of their specific health problems.

14. Mechanism for Consumer Feedback

We will make it easy for consumers to provide us with feedback or complaints concerning our health web sites.

* * *

Founding members of Hi-Ethics intend to implement these principles within six months. Contracts with a third party and a health web site in effect when these principles are adopted need not be amended, if the health web site has a good faith belief that the contract is in compliance with the principles set forth herein.

Hi-Ethics Glossary

For purposes of the Hi-Ethics Principles—

Aggregate Data means personal information or non-personal individual information collected from a group of users that has been processed so that it can no longer be used to identify a single, unique individual.

Co-branding refers to the joint branding of a web page or section of a consumer health web site between two or more corporate entities or individuals. Co-branding may involve the joint operation of services, health information content or products that appear on a consumer health web site.

Health Information Content includes information to help consumers stay well, prevent and manage disease, and make decisions related to health and health care, including information for making decisions about health-related products and health services. It may be in the form of data, text, graphics, audio or video, and may involve special software or hardware and programming enhancements that support interactivity. Health information content includes both materials authored by third parties (whether scholarly works by scientists and clinicians or interpretive articles prepared for consumers), as well as materials created specifically for use on a health web site.

Health-related Personal Information refers to personal information that is associated with health issues, categories, questions, and facts obtained as a result of the individual's responses and activities on a health web site.

Internet Health Services means the full range of services and activities available on a consumer health web site. Examples include the sale of health care products, delivery of health care services and health information, specialized health information searches, self-assessment tools and activities, bulletin boards, chat rooms with and without participation by health professionals, and opportunities for relationships and communication with health care professionals and health plans.

Non-Personal Individual Information does not include any information that would meet the definition of personal information (below), but may include information about a specific individual's characteristics, preferences, interests, experiences, and activities disclosed by the individual to the health web site or obtained through the individual's use of the health web site.

Operate refers to the degree of control a corporation or individual has over the operations of a consumer health web site. A corporation or individual operates a consumer health web site if the corporation or individual is primarily responsible for the material that appears on the site, including, but not limited to, advertising, health information content, services and products.

Opt-in means an affirmative ability for a consumer to accept terms and conditions.

Personal Information means any individually identifiable information about an individual collected online, including a first and last name, a home or other physical address, including street name and name of a city or town, an E-mail address, a telephone number, a Social Security number, or any other identifier that may permit the physical or online contacting of a specific individual.

Self-Assessment Tools refers to online forms that allow an individual to supply personal information and health-related information that cause interactive software programming using medical knowledge to reach conclusions that may be relevant to optimizing health care decisions or possible health outcomes.

Unrelated Third Party refers to a corporate entity or individual who acts on its own behalf and in its own interest and to carry out a purpose other than that for which the individual accessed the consumer health web site.

Appendix 6

Criteria for Assessing the Quality of Health Information on the Internet

Health Summit Working Group, Mitretek Systems

(Available: http://hitiweb.mitretek.org/docs/policy.html/ Accessed: October 15, 2001)

The Health Summit Working Group selected, defined, ranked and evaluated seven major criteria for assessing the quality of Internet health information: credibility, content, disclosure, links, design, interactivity, and caveats (advisories).

Criteria for Evaluating Internet Health Information

Credibility: includes the source, currency, relevance/utility, and editorial review process for the information.

Content: must be accurate and complete, and an appropriate disclaimer provided.

Disclosure: includes informing the user of the purpose of the site, as well as any profiling or collection of information associated with using the site.

Links: evaluated according to selection, architecture, content, and back linkages.

Design: encompasses accessibility, logical organization (navigability), and internal search capability.

Interactivity: includes feedback mechanisms and means for exchange of information among users.

Caveats: clarification of whether site function is to market products and services or is a primary information content provider.

Credibility

To determine the credibility of Internet health information, one must consider its source, currency, relevance/utility, and editorial review process.

Source

The source of medical information is the premier criterion for its credibility and quality. Therefore, a site should display the name and logo of the institution or organization responsible for the information, as well as the name and the title of the author(s), if relevant. Additionally, the qualifications/credentials of the organization and author(s), along with any relevant personal or financial associations or other real or potential sources of bias, should be disclosed. Disclosing sponsorship and the nature of such support can help consumers assess the motivations of the provider of the information (e.g., if the site is advertising a product or service) and any potential conflicts of interest.

Currency

The date of the original document on which the information is based and the date of posting on the Web should be provided so the user can judge the timeliness of the information. Although the date of posting does not necessarily show that the information provided is correct or up to date, it does serve as an indicator of currency.

Relevance

Users need to know the content of a site corresponds to the information it purports to offer.

Site Evaluation

In an academic community, the peer review process is used to ensure the validity and quality of the information presented in papers and reports. The general public, however, is more likely to understand a "seal of approval" from an individual or group commonly perceived as credible. Sites should indicate whether the information provided has been subjected to review, and if so, describe the process and the individuals involved.

Content

The content of health information on the Internet must be accurate and complete; an appropriate disclaimer should also be provided.

Accuracy

Given that accuracy of content is based on evidence and its verification, the site should identify the data that underlies the conclusions presented. Clinical or scientific evidence that supports a position should be clearly stated. The framework of the study should also be described in language the lay person can understand. And users need to be aware that testimonials are not evidence.

Disclaimer

A disclaimer describing the limitations, purpose, scope, authority, and currency of the information should be provided. To ensure accuracy and avoid plagiarism and copyright violation, sources of the information should also be disclosed. The disclaimer should emphasize as well that the content is general health information, not medical advice.

Completeness

The treatment of a topic should be comprehensive and balanced. Users should be wary of one-sided views with critical information missing. Pertinent facts, negative results, and a statement of any information not known about the subject should be included.

Disclosure

Websites should provide appropriate disclosures, including the purpose of the site, as well as any profiling/collecting of information associated with using the site, so users can understand the intent of the organization or individual in providing the information.

Purpose

The mission or purpose of the site should be clearly stated, and the information provided should be appropriate to that mission or purpose.

Profiling/Collection of Information

Websites request and use information for purposes of which the user may be unaware. It is critical that users be informed of the collection, use, and dissemination of any information they may be providing in visiting the site. Only then can they make an informed decision to provide the information and/or approve of its eventual use.

Links

Especially critical to the quality of an Internet site are its external links—connections to other internal pages or to external sites that form the web-like structure of information searches within and among sites. There are four criteria for evaluating the quality of links: selection, architecture, content, and back linkages.

Selection

The selection of links is made at the originating site. The person or group responsible for link selection should have the expertise and credentials to evaluate critically the appropriateness of those links. It is also important that the original and linked sites target a set of readers with similar characteristics.

Architecture

The architecture or design of pointers to linked sites is important for ease of navigation: whether there are timely escape mechanisms during side searches, whether the user can easily find his or her way backwards and forwards, and whether the structure is apparent and logical to the reader. Image-based icons and textual identifiers should be meaningful and consistent.

Content

The content of the links should be accurate, current, credible, and relevant. The content of the originating site is enhanced if it includes links to high-quality sites; on the other hand, links to poor-quality sites indicate a lower-quality originating site. Users should be alerted when they are about to view an external site. Means that can be used for this purpose include information relating to the linked source, provided before the user clicks to the site, and use of transition screens.

Back Linkages

Back linkages are links to one website from another. Many websites track and publish back linkages for the purpose of enhancing their credibility and marketability. The best way to evaluate back linkages is to examine the context in which they are used, that is, their purpose, relevance, credibility, and authority, as well as any associated bias.

Design

The design or layout of the website, including graphics and text, as well as links, is important to the effective delivery and use of any Web-based information, even though it does not affect the quality of the information per se. The design of websites can be evaluated in terms of accessibility, logical organization (navigability), and internal search capability.

Access

Websites should be accessible by the lowest-level available browser technology. Other features to improve access include options for accessing the information when multimedia browsers are not available, as well as options for enabling use by the hearing and seeing impaired.

Logical Organization (Navigability)

The best websites are clearly focused on their purpose and target audience. They are simple, internally consistent, and easy to use. Cross-references are provided to aid the user in comprehending the overall structure of the information. The composition of the information reflects an awareness of reading level, language, and the need for a balance of text and graphics, color and sound.

Internal Search Capability

An internal search engine is a highly desirable component for most websites with depth and breadth of content. The scope and function of the search engine—what it covers and how it works—should be clearly described. The search engine should be capable of searching specified content by keyword or search string and retrieving only relevant materials. It should also have a user interface that is easy to understand and use.

Interactivity

Websites should include a feedback mechanism for users to offer their comments, corrections, and criticisms, and raise questions about the information provided. This makes the website accountable to its users. If a site provides a chat room, allowing information to be exchanged among many individuals, an indication of whether a moderator is present should be provided, together with a warning that the information may not be accurate. If a moderator is present, his/her expertise and affiliations, as well as the source of his/her compensation should be identified. When a website provides an interactive service, such as tailoring information to the user based on clinical algorithms, the algorithm used should be posted, including its developer and the site's affilon with the developer.

Caveats

Sites that market services and products have different agenda then those that are primary content providers.

Summary

The Internet presents a powerful mechanism for helping users improve their health-care decision making by providing easy and rapid access, exchange, and dissemination for enormous amounts of health information. Yet users must be aware of the potential for misinformation and recognize the critical need to assess the quality of the information provided. Content providers must be encouraged to develop and post high-quality information, and policymakers and health-care professionals must be educated on this important health issue. The Health Summit Working Group has developed this set of criteria to address this critical need. These criteria are intended as a resource for users seeking health-related information on the Internet, and should aid in evaluating information to determine whether it is usable and credible.

Contact Information

For more information on the Health Summit Working Group or to join, visit our homepage—http://hitiweb.mitretek.org/hswg.

Post comments regarding the policy paper directly to http://hitiweb.mitretek.org/docs/criteria.html.

Information Quality Tool

(Available: http://hitiweb.mitretek.org/iq/ Accessed: October 15, 2001)

The IQ Tool helps you become an educated consumer by helping you ask the right questions. Based on your answers IQ will evaluate the site's strengths and weaknesses.

Mitretek Systems supports The Health Summit Working Group's "Criteria for Assessing the Quality of Health Information on the Internet." No funding is currently available to support any further development of the IQ Tool or any testing of its validity and/or reliability. Mitretek hopes that you will use the IQ Tool to help you become better educated in evaluating health web sites. If you have comments Mitretek welcomes your feedback.

Information Quality Tool Beginner's Guide

How to Use the Tool

1. You begin by selecting Begin Using IQ on IQ Tool homepage.
2. Enter the URL, the address that begins http://...., of the site that you would like to test. Select Open Page (click when arrow is over the word).
3. A second window will open. This windows will open. One window will have a question in it. The second window will open to the site you want to test. If the site does not open check the address and redo step 2.
4. Answer the questions as best you can. To answer these questions you will need to search through the site. If you are unsure you can skip the question. (this may cause the site to fail)
5. When you have answered all of the questions press the score site button.
6. You will see a report with three sections: A review of answers, the score and information on the importance of what was missing from the site.
7. The score is a grade that tells you how well a site answered the questions. It is no guarantee that a site is either good or bad.
8. The most important part of the report is the detailed information on what was missing from the site. It will tell you why the answers are important.
9. You may want to print the results for your records.
10. To test another site, go back to the IQ Tool homepage.
11. You do not have to close the window of the site that you have already evaluated. It will be updated with the new site that you enter.

General Questions

Address:

If this is promotional site selling a product, click here.
If this is promotional site selling a service, click here.

Note: Scroll down to see more questions. If a question is not applicable, leave the response blank. For quick reference help on a question, click on the button beside the question. To comment on a question, click on the button.

1. Is the author identified in the article?
   Yes   No
2. When the author refers to another source, are appropriate references provided?
   Yes   No
3. If the author is not referring to a source, does he/she clearly state that it is only his/her opinion?
   Yes   No
4. Are the site author's credentials listed?
   Yes   No
5. Does the site author's credentials relate to the knowledge of the field that is required for the site's subject discussions?
   Yes   No
6. Are the author's experiences relevant to the topic?
   Yes   No   Unknown
7. Is a means provided to contact the author directly?
   Yes   No
8. Can you determine who has paid for or sponsored this website?
   Yes   No
9. Is any financial conflict or bias explained?
   Yes   No
10. Does the site state that contributors or sponsors have no control over content?
    Yes   No
11. Is there a means to determine how current the information in the website is, for example - date of last update or posted date?
    Yes   No
12. Is the information current?
    Yes   No
13. Is the information still relevant? (for example, a new HIV treatment posted two years ago may no longer be the most appropriate treatment today)
    Yes   No
14. From your own knowledge and experience, does this site give good medical information?
    Yes   No
15. Is the medical information presented in a balanced and neutral form?
    Yes   No
16. Are the linked sites current?
    Yes   No
17. Do the linked sites give good medical information?
    Yes   No

18. If you are allowed to input information or submit queries, is a statement provided that explains whether or not this information is confidential and secure?

     Yes   No

19. Is the site easily navigable and presented in an organized manner?

     Yes   No

20. Is a search engine provided?

     Yes   No

21. Does the search engine assist you in using the site?

     Yes   No

Score Site

Appendix 7

Principles Governing AMA Publications Web Sites

(Available: http://pubs.ama-assn.org/ama_web.html/ Accessed: October 15, 2001)

The following guidelines are based on the Principles Governing AMA Web Sites, published in *JAMA* (Winker MA, Flanagin, A, Chi-Lum, B, et al. Guidelines for medical and health information sites on the Internet: principles governing AMA Web sites. *JAMA*. 2000;283:1600–1606.) The guidelines provided herein apply to all American Medical Association (AMA) Publications Web sites. These guidelines were posted March 19, 2000.

A standing committee composed of AMA staff members from the Scientific Publications and Multimedia, Publishing and Business Development, Ethical Standards, and Internet and Database Services areas developed these guidelines and will review the guidelines regularly and revise as necessary. The committee will seek review and comment from an advisory panel of individuals outside the AMA with expertise in Web-based content, advertising, privacy and confidentiality, and e-commerce. In addition, the committee seeks comments from users of the site. Please send your comments and questions to jama-comments@ama-assn.org.

I. Principles for Content

The AMA is committed to providing medical and health information of high quality via its Web sites. Visitors to AMA Publications Web sites will be given information, navigational direction, and tools needed to judge the quality, reliability, objectivity, sources, and funding of content and to make effective use of content.

Definition of Content

Content is defined as all material (including text, graphics, tables, equations, audio, and video) and menu/directional icons, bars, indicators, listings, and indexes. These principles also address functions that support content (eg, links, navigation, searches, calculations).

Site Ownership

Web site ownership, including affiliations, strategic alliances, and significant investors, should be clearly indicated on the home screen or directly accessible from a link on the home screen. Copyright ownership of specific content should be clearly indicated on screen and on items printed from the site.

Site Viewing

The site should provide information about the platform(s) and browser(s) that permit optimal viewing in a location that is easy to find.

Viewer Access, Payment, and Privacy

Information about restrictions on access to content, required registration, and password protection (if applicable) should be provided and easy to find. Information about payment (ie, subscriptions, document delivery, pay per view, etc) should be provided and easy to find. See "Principles for E-commerce" herein. Information about privacy should be provided and easy to find. See "Principles for Privacy and Confidentiality" herein."

Funding and Sponsorship

Funding or other sponsorship for any specific content should be clearly indicated and should comply with the "Principles for Advertising and Sponsorship" herein. Content should be easily distinguished from advertising as described in "Principles for Advertising and Sponsorship."

Quality of Editorial Content

Guidelines for editorial content review, posting dates, and sources were developed based on experience with the AMA Scientific Publications' sites. All AMA scientific publications and consumer site information adhere to these guidelines.

Review

Content should be reviewed for quality (including originality, accuracy, and reliability) before posting. Clinical editorial content should be reviewed by content experts not involved in creation of the content, and the content should be revised appropriately in response to such review. The method of review will be determined by individual sites. (For example, Scientific

Publications sites include peer review. Other sites rely on review by editorial boards.)

The language complexity of the content should be appropriate for the site's audience. Content should be reviewed for grammar, spelling, and composition before should be posted on the site (eg, see JAMA Author Instructions section on Peer Review).

A list of staff members and other individuals (eg, editorial board) responsible for content quality, other than anonymous peer reviewers, should be posted on the site. (eg, see JAMA's About This Journal page)

Date of Posting, Revising, and Updating and Timeliness of Editorial Content

The dates that content is posted, revised, and updated should be clearly indicated. Procedures for updating and removing time-sensitive content should be developed, implemented, and periodically reviewed to ensure that the updating and review schedule is appropriate. (For example, content can be sorted by date posted and all content older than 6 months reviewed for timeliness and accuracy.) An indication of significant revisions to any specific content should be posted and may include instructions to discard copies of versions previously printed or downloaded.

Sources of Editorial Content

Source for specific content should be clearly identified (ie, author byline or names of individual, organizational, departmental, institutional, agency, or commercial provider/producer).

Affiliations and relevant financial disclosures for authors and content producers should be clearly indicated.

Individuals who post content in online discussions, chat rooms, and e-lists should be instructed to disclose financial interests and commercial funding or affiliations related to the subject of the posted content discussion, chat, or list.

Reference material used to develop content should be cited in a manner appropriate for the site's audience.

Linking

Intrasite content links should be reviewed before posting and maintained and monitored. If links are not functional, links should be repaired in a timely manner.

External site links should be reviewed before posting and maintained and monitored. If links are not functional, these links should be repaired in a timely manner.

External links to commercial sites must comply with the "Principles for Advertising and Sponsorship."

Intersite Navigation

Sites should not prevent viewers from returning to a previous site.
Sites should not redirect the viewer to a site the viewer did not intend to visit.
Sites should not frame other sites without permission.

Downloading Files

If content can be downloaded in a portable document file (PDF) format, instructions regarding how to download the PDF file and how to obtain the necessary software should be provided and easy to find.

A link to such software should be provided. (See Frequently Asked Questions)

Navigation of Content

Features that facilitate use of the site should be provided and easy to find, and should include a site map or other site organizational guide, a help function or frequently-asked-questions page, a feedback mechanism, and customer service information (if available).

Each distinct site should provide a search engine or appropriate navigation tool to facilitate use. If the site provides a search engine, instructions specifying how to use the search function and how to conduct different types of searches may be provided.

Graphics files should include a "mouse over" indication of the graphical content. For large files, the space where the file resides should include the size of the file. As a courtesy to the viewer, when possible, when a large file can be downloaded by clicking, the viewer may be informed of the size of the file before the file begins downloading and should have the opportunity to cancel the download.

II. Principles for Advertising and Sponsorship

These principles are a portion of the Principles Governing Advertising in Publications of the American Medical Association (previously revised in May 1999). The complete set of Principles, which can be accessed by clicking on the hypertext link in this paragraph ("Principles Governing...") includes specific information for advertisers (such as types of products and services that may be advertised, advertising copy, price comparisons, and time requirements) not included herein.

These principles are applied by the AMA to ensure adherence to the highest ethical standards of advertising and to determine the eligibility of products and services for advertising on the AMA Publications Web sites. The appearance of advertising on the AMA Publications Web sites is neither a guarantee nor an endorsement by the AMA of the product, service, or company or the claims made for the product in such advertising. The fact that an advertisement for a product, service, or company has appeared on the AMA Publications Web sites shall not be referred to in collateral advertising.

As a matter of policy, the AMA will sell advertising space on its Publications Web sites when the inclusion of advertising does not interfere with the mission or objectives of the AMA or its publications.

To maintain the integrity of the AMA Publications Web sites, advertising (ie, promotional material, advertising representatives, companies, or manufacturers) cannot influence editorial decisions or editorial content (as defined in "Principles for Content"). Decisions to sell advertising space are made independently of and without information pertinent to specific editorial content. The AMA Publications Web sites' advertising sales representatives have no prior knowledge of specific editorial content before it is published. Placement of advertising adjacent to (ie, next to or within) editorial content on the same topic is prohibited (for the table of contents, a banner advertisement must not appear next to the title of a related article). Just as a print advertisement should not be placed next to an editorial page on the same topic, a digital advertisement should not be adjacent to editorial content on the same topic, either by linking or appearing adjacent in the content section of the same screen. Similarly, just as a print reader can choose to read an advertisement or skip over it, a computer user should have the option to click or not click on an advertisement. Viewers will not be sent to a commercial site unless they choose to do so by clicking on an advertisement.

The AMA, in its sole discretion, retains the right to decline any submitted advertisement or to discontinue posting of any advertisement previously accepted.

Advertising

1. Digital advertising may be placed on the AMA Publications Web site.
2. Digital advertisements must be readily distinguishable from editorial content. If the distinction is unclear, the word "advertisement" should be added.
3. Digital advertisements may appear as fixed banners or as rotating advertisements.
4. Digital advertisements may not be juxtaposed with, appear in line with, or appear adjacent to editorial content on the same topic, or be linked with editorial content on the same topic.
5. Digital advertisements that are fixed in relation to the viewer's screen or that rotate should be placed to ensure that juxtaposition (as defined in item 4 above) will not occur as screen content changes.

6. Digital banner advertisements should be limited to 1 advertisement per screen view.

7. Advertisements and promotional icons may not appear on the home page of the AMA Publications Web site (http://pubs.ama-assn.orgl) or the home pages of JAMA (http://www.jama.com) or the AMA Archives Journals (Archives home: http://pubs.ama-assn.org/archives_home.html; Archives of Dermatology: http://archderm.ama-assn.org/; Archives of Facial Plastic Surgery: http://archfaci.ama-assn.org/; Archives of Family Medicine: http://archfami.ama-assn.org/; Archives of General Psychiatry: http:// archpsyc.ama-assn.org/; Archives of Internal Medicine: http:// archinte.ama-assn.org/; Archives of Neurology: http://archneur.ama-assn.org/; Archives of Ophthalmology: http://archopht.ama-assn.org/; Archives of Otolaryngology—Head & Neck Surgery: http:// archotol.ama-assn.org/; Archives of Pediatrics & Adolescent Medicine: http://archpedi.ama-assn.org/; and Archives of Surgery: http:// archsurg.ama-assn.org/).

8. AMA, JAMA, and Archives Journals logos may not appear on commercial Web sites as a logo or in any other form without prior written approval by the individuals responsible for the respective areas within AMA.

9. Advertisements may link to additional promotional content that resides on the AMA Publications Web site.

10. Advertisements may link off-site to a commercial Web site, provided that the viewer is clearly informed with a buffer page that to proceed by clicking would mean the viewer would leave the AMA Publications Web site and that the AMA Publications Web site does not vouch for or assume any responsibility for any material contained on the Web site to which it links. The buffer page will display the following statement:

You are leaving the AMA Publications Web site. If you wish to link to a Web site maintained by [company name], please click below. If you do not wish to leave the AMA Publications Web site, please click on the "back" button of your browser to return to the site. The AMA does not assume responsibility for content of other Web sites.

The AMA will not link to Web sites that frame the AMA Publications Web site content without express permission of the AMA; prevent the viewer from returning to the AMA Publications Web site or other previously viewed screens, such as by disabling the viewer's "back" button; or redirect the viewer to a Web site the viewer did not intend to visit.

The AMA reserves the right to not link to or to remove links to other Web sites.

11. Methods of corporate funding should be described in the Publications Web site's information about advertising or the digital rate card.

Sponsorship

1. All financial or material support for electronic collections of articles, Publications Web site content, and other types of online products (such

as JAMA Information Centers Web sites, online databases, or material on CD-ROM) will be acknowledged and clearly indicated on the home screen or via a link from the home screen.

2. Acknowledgment of support will appear on the home page, on the running foot of all pages, on any packaging and collateral material included (eg, CD-ROM jewel case and companion print insert), and on any materials used to publicize the online product. Content accessed through the site that does not reside on the site (eg, abstracts or articles from another site) will not include sponsorship information.

3. These acknowledgments will not make any claim for any supporting company product(s). The final wording and positioning of the acknowledgment will be determined by the AMA. The wording will be similar to "Produced by [AMA publication] with support from [Company]."

4. The home page acknowledgment of digital products may be linked to an on-site "About [Company]" page or may link to the company's Web site through the intervening buffer page referred to in "Advertising," item 10.

5. The "About [Company]" page may be linked to other on-site pages provided by the supporting company. These pages must be readily distinguishable from editorial content, must be clearly labeled as provided by the supporting company, and must not be linked to related AMA editorial content.

6. The running foot acknowledgment will not be linked to any other materials deemed necessary by AMA Publications.

7. AMA, JAMA, and Archives Journals logos may not appear on the sponsoring company Web site as a logo or in any other form without prior written approval by the individuals responsible for the respective areas within the AMA.

III. Principles for Privacy and Confidentiality

The following principles reflect the AMA's commitment to maintain the Publications Web site visitor's rights to privacy and the confidentiality of personal information. In this context, privacy refers to the right of the individual site visitor to choose whether to allow personal information to be collected, by the host site (in this case, the AMA) or by third parties, and to know what type of information is collected and how that information is used. Confidentiality is the right of an individual to not have personally identifiable medical or other information disclosed to others without that individual's express informed consent.

The Internet has the potential to allow information about Publications Web site use to be tracked in aggregate (which can help site developers understand site use and improve the experience of the viewer) and at the individual user level. Individual user information can improve the visitor's experience of the site by permitting personalization of the site related to the individual's particular interests or concerns. However, tracking of personal medical and health information (ie, medical conditions, health-seeking behaviors and questions, and requests about drug therapies or medical devices or information pertaining to them) could breach an individual's personal privacy and reveal an individual's health data.

Thus, health and medical Publications Web sites have a particular obligation to protect the privacy and confidentiality of individuals. Patients and individuals with interest in particular medical conditions should feel confident in obtaining information and using resources on the site, without concern that such use will be identified with them without their permission. The AMA believes that all site visitors should have the opportunity to opt in or out of allowing personal information to be tracked. In addition, the AMA takes extensive measures to ensure the safety and security of its Publications Web site servers and to guard against divulging private information. The AMA believes that Publications Web site visitors should know who (eg, the site organization or third party) is tracking personal information and the types of personal information that are tracked and should have the right to opt out of such information being collected at any time.

The AMA Publications Web sites do not collect personal information of viewers other than e-mail addresses of individuals wishing to receive e-mail alerts of AMA Publications Tables of Contents or individuals who purchase online subscriptions. Such information is not used for any purpose other than the service the individual requests.

Protection of patients' rights to confidentiality is fundamental to medical publishing. Health care professionals must adhere to privacy and confidentiality principles to legally and ethically share important information about medical conditions of individual patients. The sharing of such information may improve clinical care for the individual or improve the general state of knowledge about medical and health care through medical research. Medical publications, whether in print or online, must not reveal identifiable information about an individual without that person's express informed consent. These principles apply to information in medical publications (eg, JAMA) as well as less formal venues used by health care professionals, such as online discussion groups, chat rooms, and e-lists.

Privacy

1. A link to the privacy policy of the Publications Web site should be provided on the home page or the site navigational bar and should be easily accessible to the user. The Publications Web site should adhere to the privacy principles posted.

2. Individuals responsible for Web sites that post advertising should be aware of current technology and access possessed by third parties that post or link

to advertisements. Web sites should ensure that the technology and access used by third parties adheres to the Web site's privacy policies.

3. The site should not collect name, e-mail address, or any other personal information unless voluntarily provided by the visitor after the visitor is informed about the potential use of such information.

4. The process of opting in to any functionality that includes collection of personal information should include an explicit notice that personal information will be saved, with explanation of how the information will be used and by whom. The opt-in statement should not be embedded in a lengthy document and should be explicit and clear to the viewer.

5. Collection, retention, and use of nonmedical personal information about site visitors may be offered to viewers when the AMA believes that such information would be useful in providing site visitors with products, services, and other opportunities, provided such use adheres to these principles and is within bounds of current regulations and law (http://www.ftc.gov/privacy/index.html). Individuals may agree to have such nonmedical personal information collected or may choose not to, with the understanding that opting out of having such information collected prevents the site from being tailored to their particular needs and interests. Such information will not include personal health information, such as any information about medical conditions or medications purchased.

6. Names and e-mail addresses of site visitors should not be provided or released to a third party without the site visitor's express permission.

7. E-mail information, personal information about specific visitor's access and navigation, and information volunteered by site visitors, such as survey information and site registration information, may be used by the site owner to improve the site but should not be shared with or sold to other organizations for commercial purposes without the site visitor's express permission.

8. The AMA will use e-mail addresses voluntarily provided by sitevisitors to notify them about updates, products, services, activities, or upcoming events. Site visitors who do not wish to receive such notifications via e-mail should be able to opt out of receiving such information at any time.

9. The AMA has licensed its physician and medical student list to third parties for more than 50 years. This information is licensed to database licensees under strict guidelines. The names and addresses of physicians in the AMA Physician Masterfile are made available only for communications that are germane to the practice of medicine or of interest to physicians or medical students as consumers. E-mail addresses are excluded from such licensing agreements.

10. Nonidentifiable Publications Web site visitor data may be collected and used in aggregate to help shape and direct the creation and maintenance of content and to determine the type of advertisement to be seen by site visitors while on the AMA site.

11. The AMA will not collect and will not allow third parties to collect personal medical information (medical conditions, health-seeking behaviors and questions, and use of or requests for information about drugs, therapies, or medical devices) without the express consent of the site visitor after explanation of the potential uses of such information.

12 A cookie is a small file stored on the site user's computer or Web server and is used to aid Web page navigation. Two types of cookies are commonly used. A session cookie is a temporary file created whenever a Web site is accessed and is self-terminated based either on an expiration date (eg, 3 hours from creation of the cookie) or by closing the Web browser. A persistent cookie is a permanent file and must be deleted manually. Cookies referred to in the context of these Guidelines are persistent cookies. A cookie function may be used on the site to track visitor practices to help determine which site features and services are most important and guide editorial direction. The cookie makes it possible for the user to access the site without requiring entry of a user name or password, allows the user to view different restricted areas of the site without reregistering, allows the user to personalize the site for future use, and permits the user to make subsequent purchases without reentering credit card information. Users who do not desire the functionality created by the cookie should have the option to disable the cookie function, either by indicating when asked that they do not wish to have a cookie created or by disabling the cookie function on their browser. Individuals should be able to opt out of cookie functions that permit tracking of personal information at any time. At this time, the AMA Publications Web sites do not use persistent cookies. Users will be notified if and when AMA Publications Web sites begin using persistent cookies, as specified in these guidelines.

13. E-mail messages sent to a Web site may not be secure. Site visitors should be discouraged from sending confidential information by e-mail. Site visitors sending e-mail accept the risk that a third party may intercept e-mail messages.

14. Market research conducted by the site or its agent to enhance the site should be clearly identified as such.

15. E-mail alerts and newsletters should contain an "unsubscribe" option.

Confidentiality

Content published within the AMA Publications Web sites that includes patient information should adhere to the patient privacy and anonymity principles followed by JAMA and the Archives Journals, which are based on the recommendations of the International Committee of Medical Journal Editors (http://jama.ama-assn.org/info/auinst_req.html). These principles apply equally to formal medical publications and the informal interactive communication permitted by the Web, including online discussion groups, chat rooms, or e-lists.

Patients should be aware when they provide information about their individual medical conditions in the context of such discussions that information may be linked with a personal identifier. However, AMA Publications Web sites will not collect information about individual medical conditions without the express permission of the site visitor. Physicians and other health care professionals should be aware that any patient information reported in the context of such venues must adhere to the confidentiality principles listed herein. Moderators of such sessions should make every effort to ensure that listed material adheres to the principles stated herein and, when in doubt, should query the individual providing the information. If the individual is a patient providing such information, the moderator should query the patient as to whether the patient intends for the sensitive medical information to be revealed. If the individual providing the information is a health care professional, the moderator should query the professional as to whether the patient reported has provided informed consent and state so.

Patients have a right to privacy that should not be infringed without express informed consent. Identifying patient information should not be published in print or online descriptions, photographs, or pedigrees (illustrations of how a disease is expressed within an extended family for purposes of determining possible inheritance) unless the information is essential for scientific purposes and the patient (or parent or guardian) gives express informed consent for publication.

Identifying details should be omitted if they are not essential, but patient data should never be altered or falsified in an attempt to attain anonymity. Complete anonymity is difficult to achieve, and informed consent should be obtained if there is any possibility as to whether identifiable information may be disclosed.

When express informed consent has been obtained, it should be indicated in the posted Publications Web content.

IV. Principles for e-Commerce

The AMA e-commerce principles are intended to ensure that users and purchasers of information, products, and services on the site will have access to secure, efficient transactions for online and remote customer fulfillment. All such transactions should adhere to the AMA "Principles for Privacy and Confidentiality." The AMA Publications Websites do not currently use e-commerce functions.

1. A link or reference to the site's policies on privacy should be clearly visible.
2. The security software and encryption protocol used on the site for financial
3. Users should be able to select whether or not the Web host will retain the user name and password (ie, disable cookie function, as described in

"Principles for Privacy and Confidentiality"). Users should be able to opt in or opt out of functions that track personal information at any time.

4. A link or reference to customer service contact information (e-mail, telephone, fax, mail), including hours of operation and time zone, should be clearly visible.

5. The terms of use for e-commerce should require a deliberate selection (accept/not accept).

6. Users should be able to review transaction information prior to execution (information, products, and services listed; prices; totals; shipping and handling expenses).

7. As a courtesy, following execution of the transaction, users should be provided, on a page or by e-mail, purchase information (see item 6 above) as well as shipping tracking number, if appropriate.

8. Users will be notified on-screen when entering or leaving a secure site and will have the option to proceed or remain on the current site.

9. If a user's browser does not support a secure connection, no financial transactions will be permitted over the Internet.

10. Response times for feedback and fulfillment should be clearly stated.

11. Products and services will not be endorsed or cobranded by the AMA or AMA publications. Any product promotions must adhere to the "Principles for Advertising and Sponsorship."

Index

Health Informatics Series
(formerly Computers in Health Care)